Springer Undergraduate Mathematics Series

Advisory Board

M.A.J. Chaplain *University of Dundee*
K. Erdmann *Oxford University*
A.MacIntyre *University of London*
L.C.G. Rogers *Cambridge University*
E. Süli *Oxford University*
J.F. Toland *University of Bath*

Other books in this series

A First Course in Discrete Mathematics *I. Anderson*
Analytic Methods for Partial Differential Equations *G. Evans, J. Blackledge, P. Yardley*
Applied Geometry for Computer Graphics and CAD, Second Edition *D. Marsh*
Basic Linear Algebra, Second Edition *T.S. Blyth and E.F. Robertson*
Basic Stochastic Processes *Z. Brzeźniak and T. Zastawniak*
Calculus of One Variable *K.E. Hirst*
Complex Analysis *J.M. Howie*
Elementary Differential Geometry *A. Pressley*
Elementary Number Theory *G.A. Jones and J.M. Jones*
Elements of Abstract Analysis *M. Ó Searcóid*
Elements of Logic via Numbers and Sets *D.L. Johnson*
Essential Mathematical Biology *N.F. Britton*
Essential Topology *M.D. Crossley*
Fields, Flows and Waves: An Introduction to Continuum Models *D.F. Parker*
Further Linear Algebra *T.S. Blyth and E.F. Robertson*
Geometry *R. Fenn*
Groups, Rings and Fields *D.A.R. Wallace*
Hyperbolic Geometry, Second Edition *J.W. Anderson*
Information and Coding Theory *G.A. Jones and J.M. Jones*
Introduction to Laplace Transforms and Fourier Series *P.P.G. Dyke*
Introduction to Ring Theory *P.M. Cohn*
Introductory Mathematics: Algebra and Analysis *G. Smith*
Linear Functional Analysis *B.P. Rynne and M.A. Youngson*
Mathematics for Finance: An Introduction to Financial Engineering *M. Capiński and
    T. Zastawniak*
Matrix Groups: An Introduction to Lie Group Theory *A. Baker*
Measure, Integral and Probability, Second Edition *M. Capiński and E. Kopp*
Multivariate Calculus and Geometry, Second Edition *S. Dineen*
Numerical Methods for Partial Differential Equations *G. Evans, J. Blackledge, P.Yardley*
Probability Models *J.Haigh*
Real Analysis *J.M. Howie*
Sets, Logic and Categories *P. Cameron*
Special Relativity *N.M.J. Woodhouse*
Symmetries *D.L. Johnson*
Topics in Group Theory *G. Smith and O. Tabachnikova*
Vector Calculus *P.C. Matthews*

John M. Howie

Fields and Galois Theory

With 22 Figures

 Springer

John M. Howie, CBE, MA, DPhil, DSc, Hon D. Univ., FRSE
School of Mathematics and Statistics
University of St Andrews
North Haugh
St Andrews
Fife
KY16 9SS
UK

Cover illustration elements reproduced by kind permission of:
Aptech Systems, Inc., Publishers of the GAUSS Mathematical and Statistical System, 23804 S.E. Kent-Kangley Road, Maple Valley, WA 98038, USA. Tel: (206) 432 -7855 Fax (206) 432 -7832 email: info@aptech.com URL: www.aptech.com.
American Statistical Association: Chance Vol 8 No 1, 1995 article by KS and KW Heiner 'Tree Rings of the Northern Shawangunks' page 32 fig 2.
Springer-Verlag: Mathematica in Education and Research Vol 4 Issue 3 1995 article by Roman E Maeder, Beatrice Amrhein and Oliver Gloor 'Illustrated Mathematics: Visualization of Mathematical Objects' page 9 fig 11, originally published as a CD ROM 'Illustrated Mathematics' by TELOS: ISBN 0-387-14222-3, German edition by Birkhauser: ISBN 3-7643-5100-4.
Mathematica in Education and Research Vol 4 Issue 3 1995 article by Richard J Gaylord and Kazume Nishidate 'Traffic Engineering with Cellular Automata' page 35 fig 2. Mathematica in Education and Research Vol 5 Issue 2 1996 article by Michael Trott 'The Implicitization of a Trefoil Knot' page 14.
Mathematica in Education and Research Vol 5 Issue 2 1996 article by Lee de Cola 'Coins, Trees, Bars and Bells: Simulation of the Binomial Process' page 19 fig 3. Mathematica in Education and Research Vol 5 Issue 2 1996 article by Richard Gaylord and Kazume Nishidate 'Contagious Spreading' page 33 fig 1. Mathematica in Education and Research Vol 5 Issue 2 1996 article by Joe Buhler and Stan Wagon 'Secrets of the Madelung Constant' page 50 fig 1.

Mathematics Subject Classification (2000): 12F10; 12-01

British Library Cataloguing in Publication Data
Howie, John M. (John Mackintosh)
    Fields and Galois theory. - (Springer undergraduate mathematics series)
    1. Algebraic fields 2. Galois theory
    I. Title
    512.7' 4
ISBN-10: 1852339861

Library of Congress Control Number: 2005929862

Springer Undergraduate Mathematics Series ISSN 1615-2085
ISBN-10: 1-85233-986-1   e-ISBN 1-84628-181-4   Printed on acid-free paper
ISBN-13: 978-1-85233-986-9

Printed in the United States of America   (HAM)

9 8 7 6 5 4 3 2 1

Springer Science+Business Media
springeronline.com

To Dorothy, Anne,
Catriona, Sarah, Karen and Fiona,
my "monstrous regiment of women",
with much love

Preface

Fields are sets in which all four of the rational operations, memorably described by the mathematician Lewis Carroll as "perdition, distraction, uglification and derision", can be carried out. They are assuredly the most natural of algebraic objects, since most of mathematics takes place in one field or another, usually the rational field $\mathbb{Q}$, or the real field $\mathbb{R}$, or the complex field $\mathbb{C}$. This book sets out to exhibit the ways in which a systematic study of fields, while interesting in its own right, also throws light on several aspects of classical mathematics, notably on ancient geometrical problems such as "squaring the circle", and on the solution of polynomial equations.

The treatment is unashamedly unhistorical. When Galois and Abel demonstrated that a solution by radicals of a quintic equation is not possible, they dealt with permutations of roots. From sets of permutations closed under composition came the idea of a permutation group, and only later the idea of an abstract group. In solving a long-standing problem of classical algebra, they laid the foundations of modern abstract algebra. It is surely reasonable now to suppose that anyone setting out to study Galois theory will have a significant experience of the language and concepts of abstract algebra, and assuredly one can use this language to present the arguments more coherently and concisely than was possible for Galois (who described his own manuscript as *ce gâchis*[1]!) I hope that I have done so, but the arguments in Chapters 7 and 8 still require concentration and determination on the part of the reader.

Again, on this same assumption (that my readers have had some exposure to abstract algebra), I have chosen in Chapter 2 to examine the properties and interconnections of euclidean domains, principal ideal domains and unique factorisation domains in abstract terms before applying them to the crucial

---

[1] "this mess".

ring of polynomials over a field.

All too often mathematics is presented in such a way as to suggest that it was engraved in pre-history on tablets of stone. The footnotes with the names and dates of the mathematicians who created this area of algebra are intended to emphasise that mathematics was and is created by real people. Foremost among the people whose work features in this book are two heroic and tragic figures. The first, a Norwegian, is Niels Henrik Abel, who died of tuberculosis at the age of 26; the other, from France, is Evariste Galois, who was killed in a duel at the age of 20. Information on all these people and their achievements can be found on the St Andrews website `www-history.mcs.st-and.ac.uk/history/`.

The book contains many worked examples, as well as more than 100 exercises, for which solutions are provided at the end of the book.

It is now several years since I retired from the University of St Andrews, and I am most grateful to the university, and especially to the School of Mathematics and Statistics, for their generosity in continuing to give me access to a desk and a computer. Special thanks are due to Peter Lindsay, whose answers to stupid questions on computer technology were unfailingly helpful and polite. I am grateful also to my colleague Sophie Huczynka and to Fiona Brunk, a final-year undergraduate, for drawing attention to mistakes and imperfections in a draft version. The responsibility for any inaccuracies that remain is mine alone.

John M. Howie

University of St Andrews

May, 2005

Contents

1
Rings and Fields

1.1 Definitions and Basic Properties

Although my assumption in writing this book is that my readers have some knowledge of abstract algebra, a few reminders of basic definitions may be necessary, and have the added advantage of establishing the notations and conventions I shall use throughout the book. Introductory texts in abstract algebra (see [13], for example) are often titled or subtitled "Groups, Rings and Fields", with fields playing only a minor part. Yet the theory of fields, through which both geometry and the classical theory of equations are illuminated by abstract algebra, contains some of the deepest and most remarkable insights in all mathematics. The hero of the narrative ahead is Evariste Galois,[1] who died in a duel before his twenty-first birthday.

A **ring** $R = (R, +, .)$ is a non-empty set R furnished with two binary operations $+$ (called addition) and $.$ (called multiplication) with the following properties. (Under the usual convention the dot for multiplication is omitted.)

(R1) *the associative law for addition:*

$$(a + b) + c = a + (b + c) \quad (a, b, c, \in R);$$

(R2) *the commutative law for addition:*

$$a + b = b + a \quad (a, b \in R);$$

---

[1] Evariste Galois, 1811–1832.

(R3) *the existence of* 0: there exists 0 in R such that, for all a in R,

$$a + 0 = a\,;$$

(R4) *the existence of negatives:* for all a in R there exists $-a$ in R such that

$$a + (-a) = 0\,;$$

(R5) *the associative law for multiplication:*

$$(ab)c = a(bc) \quad (a, b, c \in R)\,;$$

(R6) *the distributive laws:*

$$a(b + c) = ab + ac\,, \ (a + b)c = ac + bc \quad (a, b, c \in R)\,.$$

We shall be concerned only with **commutative rings**, which have the following extra property.

(R7) *the commutative law for multiplication:*

$$ab = ba \quad (a, b \in R)\,.$$

A **ring with unity** R has the properties (R1) – (R6), together with the following property.

(R8) *the existence of* 1: there exists $1 \neq 0$ in R such that, for all a in R,

$$a1 = 1a = a\,.$$

The element 1 is called the **unity element**, or the (multiplicative) **identity** of R.

A commutative ring R with unity is called an **integral domain** or, if the context allows, just a **domain**, if it has the following property.

(R9) *cancellation:* for all a, b, c in R, with $c \neq 0$,

$$ca = cb \ \Rightarrow \ a = b\,.$$

A commutative ring R with unity is called a **field** if it has the following property.

(R10) *the existence of inverses:* for all $a \neq 0$ in R there exists a^{-1} in R such that

$$aa^{-1} = 1\,.$$

We frequently wish to denote a^{-1} by $1/a$.

It is easy to see that (R10) implies (R9). The converse implication, however, is not true: the ring $\mathbb{Z}$ of integers is an obvious example. It is worth noting also that (R9) is equivalent to

(R9)' *no divisors of zero:* for all a, b in R,

$$ab = 0 \implies a = 0 \text{ or } b = 0\,.$$

(See Exercise 1.4.)

It is useful also at this stage to remind ourselves of the definition of a group. A **group** $G = (G,.)$ is a non-empty set furnished with a binary operation . (usually omitted) with the following properties.

(G1) *the associative law*:

$$(ab)c = a(bc) \quad (a, b, c \in G)\,;$$

(G2) *the existence of an identity element*: there exists e in G such that, for all a in G,

$$ea = a\,;$$

(G3) *the existence of inverses*: for all a in G there exists a^{-1} in G such that

$$a^{-1}a = e\,.$$

An **abelian**[2] group has the extra property

(G4) *the commutative law*:

$$ab = ba \quad (a, b \in G)\,.$$

Remark 1.1

If $(R, +, .)$ is a ring, then $(R, +)$ is an abelian group. If $(K, +, .)$ is a field and $K^* = K \setminus \{0\}$, then $(K^*, .)$ is an abelian group.

Let R be a commutative ring with unity, and let

$$U = \{u \in R \,:\, (\exists v \in R)\, uv = 1\}\,.$$

It is easy to verify that U is an abelian group with respect to multiplication in R. We say that U is the **group of units** of the ring R. If a, b in R are such that $a = ub$ for some u in U, we say that a and b are **associates**, and write $a \sim b$. For example, in the ring $\mathbb{Z}$ the group of units is $\{1, -1\}$, and $a \sim -a$ for all a in $\mathbb{Z}$.

---

[2] Niels Henrik Abel, 1802–1829.

Example 1.2

Show that $R = \{a + b\sqrt{2} : a, b \in \mathbb{Z}\}$ forms a commutative ring with unity with respect to the addition and multiplication in $\mathbb{R}$. Show that the group of units of R is infinite.

Solution

It is clear that

$$(a + b\sqrt{2}) + (c + d\sqrt{2}) = (a + c) + (b + d)\sqrt{2} \in R$$

and

$$(a + b\sqrt{2})(c + d\sqrt{2}) = (ac + 2bd) + (ad + bc)\sqrt{2} \in R.$$

Since R is a subset of $\mathbb{R}$, the properties (R1), (R2), (R5), (R6) and (R7) are automatically satisfied. The ring also has the properties (R3), (R4) and (R8), since the zero element is $0 + 0\sqrt{2}$, the negative of $a + b\sqrt{2}$ is $(-a) + (-b)\sqrt{2}$, and the unity element is $1 + 0\sqrt{2}$. The element $1 + \sqrt{2}$ is in the group of units, since $(1 + \sqrt{2})(-1 + \sqrt{2}) = 1$. The powers of this element are all distinct, since $1 + \sqrt{2} > 1$, and so

$$1 + \sqrt{2} < (1 + \sqrt{2})^2 < (1 + \sqrt{2})^3 < \cdots.$$

All these powers are in the group of units, which is therefore infinite.

The group of units is in fact $\{a + b\sqrt{2} : a, b \in \mathbb{Z}, |a^2 - 2b^2| = 1\}$. $\qquad\square$

Remark 1.3

The group of units of a field K is the group K^* of all non-zero elements of K.

In a field, every non-zero element divides every other, but in an integral domain D the notion of divisibility plays a very significant role. If $a \in D \setminus \{0\}$ and $b \in D$, we say that a **divides** b, or that a is a **divisor** of b, or that a is a **factor** of b, if there exists z in D such that $az = b$. We write $a \mid b$, and occasionally write $a \nmid b$ if a does not divide b. We say that a is a **proper divisor**, or a **proper factor**, of b, or that a **properly divides** b, if z is not a unit. Equivalently, a is a proper divisor of b if and only if $a \mid b$ and $b \nmid a$.

EXERCISES

1.1. Many of the standard techniques of classical algebra are consequences of the axioms of a ring. The exceptions are those depending on commutativity of multiplication (R7) and divisibility (R10). Let R be a ring.

(i) Show that, for all a in R,

$$a0 = 0a = 0\,.$$

(ii) Show that, for all a, b in R,

$$a(-b) = (-a)b = -ab\,, \quad (-a)(-b) = ab\,.$$

1.2. What difference does it make if the stipulation that $1 \neq 0$ is omitted from Axiom (R7)?

1.3. Axiom (R7) ensures that a field has at least two elements. Show that there exists a field with exactly two elements.

1.4. Prove the equivalence of (R9) and (R9)$'$.

1.5. Show that every finite integral domain is a field.

1.6. Show that $\sim$, as defined in the text, is an **equivalence relation**. That is, show that, for all a, b, c in a commutative ring R with unity,

(i) $a \sim a$ (*the reflexive property*);

(ii) $a \sim b \Rightarrow b \sim a$ (*the symmetric property*);

(iii) $a \sim b$ and $b \sim c \Rightarrow a \sim c$ (*the transitive property*).

1.7. Let $i = \sqrt{-1}$. Show that, by contrast with Example 1.2, the ring $R = \{a + bi\sqrt{2} : a, b \in \mathbb{Z}\}$ has group of units $\{1, -1\}$.

1.8. Let D be an integral domain. Show that, for all a, b in $D \setminus \{0\}$:

(i) $a \mid a$ (*the reflexive property*);

(ii) $a \mid b$ and $b \mid c \Rightarrow a \mid c$ (*the transitive property*);

(iii) $a \mid b$ and $b \mid a \Rightarrow a \sim b$.

1.2 Subrings, Ideals and Homomorphisms

Much of the material in this section can be applied, with occasional modifications, to rings in general, but we shall suppose, without explicit mention, that all our rings are commutative. We shall use standard algebraic shorthands: in particular, we write $a - b$ instead of $a + (-b)$.

A **subring** U of a ring R is a non-empty subset of R with the property that, for all a, b in R,

$$a, b \in U \implies a - b, \ ab \in U. \tag{1.1}$$

Equivalently, $U (\neq \emptyset)$ is a subring if, for all a, b in R,

$$a, b \in U \implies a + b, \ ab \in U, \quad a \in U \implies -a \in U. \tag{1.2}$$

(See Exercise 1.2.)

It is easy to see that $0 \in U$: simply choose a from the non-empty set U, and deduce from (1.1) that $0 = a - a \in U$.

A **subfield** of a field K is a subring which is a field. Equivalently, it is a subset E of K, containing at least two elements, such that

$$a, b \in E \implies a - b \in E, \quad a \in E, \ b \in E \setminus \{0\} \implies ab^{-1} \in E. \tag{1.3}$$

Again, we may replace the second implication of (1.3) by the two implications

$$a, b \in E \implies ab \in E, \quad a \in E \setminus \{0\} \implies a^{-1} \in E. \tag{1.4}$$

If $E \subset K$ we say that E is a **proper** subfield of K.

An **ideal** of R is a non-empty subset I of R with the properties

$$a, b \in I \implies a - b \in I, \quad a \in I \text{ and } r \in R \implies ra \in I. \tag{1.5}$$

An ideal is certainly a subring, but not every subring is an ideal: the subring $\mathbb{Z}$ of the field $\mathbb{Q}$ of rational numbers provides an example. Among the ideals of R are $\{0\}$ and R. An ideal I such that $\{0\} \subset I \subset R$ is called **proper**.

Theorem 1.4

Let $A = \{a_1, a_2, \ldots, a_n\}$ be a finite subset of a commutative ring R. Then the set

$$Ra_1 + Ra_2 + \cdots + Ra_n \ (= \{x_1 a_1 + x_2 a_2 + \cdots + x_n a_n : x_1, x_2, \ldots, x_n \in R\})$$

is the smallest ideal of R containing A.

Proof

The set $Ra_1 + Ra_2 + \cdots + Ra_n$ is certainly an ideal, since, for all

$$x_1, x_2, \ldots, x_n, y_1, y_2, \ldots, y_n$$

in R,

$$(x_1a_1 + x_2a_2 + \cdots + x_na_n) - (y_1a_1 + y_2a_2 + \cdots + y_na_n)$$
$$= (x_1 - y_1)a_1 + (x_2 - y_2)a_2 + \cdots + (x_n - y_n)a_n$$
$$\in Ra_1 + Ra_2 + \cdots + Ra_n\,;$$

and, for all r in R,

$$r(x_1a_1 + x_2a_2 + \cdots + x_na_n) = (rx_1)a_1 + (rx_2)a_2 + \cdots + (rx_n)a_n$$
$$\in Ra_1 + Ra_2 + \cdots + Ra_n\,.$$

It is clear that every ideal I containing $\{a_1, a_2, \ldots, a_n\}$ contains the element $x_1a_1 + x_2a_2 + \cdots + x_na_n$ for every choice of $x_1, x_2, \ldots, x_n$ in R, and so $Ra_1 + Ra_2 + \cdots + Ra_n \subseteq I$. $\qquad\square$

We refer to $Ra_1 + Ra_2 + \cdots + Ra_n$ as the **ideal generated by** $a_1, a_2, \ldots, a_n$, and frequently write it as $\langle a_1, a_2, \ldots, a_n \rangle$. Of special interest is the case where the ideal is generated by a single element a in R; we say that $Ra = \langle a \rangle$ is a **principal ideal**.

There is a close connection between ideals and divisibility:

Theorem 1.5

Let D be an integral domain with group of units U, and let $a, b \in D \setminus \{0\}$. Then:

(i) $\langle a \rangle \subseteq \langle b \rangle$ if and only if $b \mid a$;

(ii) $\langle a \rangle = \langle b \rangle$ if and only if $a \sim b$;

(iii) $\langle a \rangle = D$ if and only if $a \in U$.

Proof

(i) Suppose first that $b \mid a$. Then $a = zb$ for some z in D, and so

$$\langle a \rangle = Da = Dzb \subseteq Db = \langle b \rangle\,.$$

Conversely, suppose that $\langle a \rangle \subseteq \langle b \rangle$. Then there exists z in D such that $a = zb$, and so $b \mid a$.

(ii) Suppose first that $a \sim b$. Then there exists u in U such that $a = ub$ and $b = u^{-1}a$. Thus $b \mid a$ and $a \mid b$ and so, by (i), $\langle a \rangle = \langle b \rangle$.

Conversely, suppose that $\langle a \rangle = \langle b \rangle$. Then there exist u, v in D such that $a = ub$, $b = va$. Hence $(uv)a = u(va) = ub = a = 1a$, and so, by cancellation, $uv = 1$. Thus u and v are units, and so $a \sim b$.

(iii) It is clear that $\langle 1 \rangle = D$. Hence, by (ii), $\langle a \rangle = D$ if and only if $a \sim 1$, that is, if and only if a is a unit. $\square$

A **homomorphism** from a ring R into a ring S is a mapping $\varphi : R \to S$ with the properties

$$\varphi(a + b) = \varphi(a) + \varphi(b), \quad \varphi(ab) = \varphi(a)\varphi(b). \qquad (1.6)$$

Among the homomorphisms from R into S is the **zero mapping** ζ given by

$$\zeta(a) = 0 \ (a \in R). \qquad (1.7)$$

While some of the theorems we establish will apply to all homomorphisms, including ζ, others will apply only to **non-zero** homomorphisms.

Some elementary properties of ring homomorphisms are gathered together in the following theorem:

Theorem 1.6

Let R, S be rings, with zero elements 0_R, 0_S, respectively, and let $\varphi : R \to S$ be a homomorphism. Then,

(i) $\varphi(0_R) = 0_S$;

(ii) $\varphi(-r) = -\varphi(r)$ for all r in R;

(iii) $\varphi(R)$ is a subring of S.

Proof

(i) Since

$$\varphi(a) + \varphi(0_R) = \varphi(a + 0_R) = \varphi(a),$$

we can deduce that

$$\varphi(0_R) = 0_S + \varphi(0_R) = -\varphi(a) + \varphi(a) + \varphi(0_R) = -\varphi(a) + \varphi(a) = 0_S. \quad (1.8)$$

(ii) Since, for all r in R,

$$\varphi(r) + \varphi(-r) = \varphi(r + (-r)) = \varphi(0_R) = 0_S = \varphi(r) + (-\varphi(r)),$$

it follows that

$$\varphi(-r) = -\varphi(r). \qquad (1.9)$$

(iii) Let $\varphi(a)$, $\varphi(b)$ be arbitrary elements of $\varphi(R)$, with $a, b \in R$. Then

$$\varphi(a)\varphi(b) = \varphi(ab) \in \varphi(R)$$

and, by virtue of (1.9),

$$\varphi(a) - \varphi(b) = \varphi(a) + \varphi(-b) = \varphi\big(a + (-b)\big) \in \varphi(R).$$

Thus $\varphi(R)$ is a subring. $\qquad\qquad\square$

The following corollary is an immediate consequence of the above proof:

Corollary 1.7

If $\varphi : R \to S$ is a ring homomorphism and $a, b \in R$, then $\varphi(a-b) = \varphi(a) - \varphi(b)$.

Let $\varphi : R \to S$ be a homomorphism. If φ is one-to-one, we call it a **monomorphism**, or an **embedding**, and if φ is also onto we call it an **isomorphism**. We say that the rings R and S are **isomorphic** (to each other) and write $R \simeq S$. For example, the ring $R = \{m + n\sqrt{2} : m, n \in \mathbb{Z}\}$ is isomorphic to the ring

$$S = \left\{ \begin{pmatrix} m & n \\ 2n & m \end{pmatrix} : m, n \in \mathbb{Z} \right\} \tag{1.10}$$

with the operations of matrix addition and multiplication, the isomorphism being

$$\varphi : m + n\sqrt{2} \mapsto \begin{pmatrix} m & n \\ 2n & m \end{pmatrix}.$$

We shall eventually be interested in the case where the rings R and S coincide: an isomorphism from R onto itself is called an **automorphism**.

If $\varphi : R \to S$ is a monomorphism, then the subring $\varphi(R)$ of S is isomorphic to R. Since the rings R and $\varphi(R)$ are abstractly identical, we often wish to identify $\varphi(R)$ with R and regard R itself as a subring of S. For example, if S is the ring defined by (1.10), there is a monomorphism $\theta : \mathbb{Z} \to R$ given by

$$\theta(m) = \begin{pmatrix} m & 0 \\ 0 & m \end{pmatrix} \quad (m \in \mathbb{Z}),$$

and the identification of the integer m with the 2×2 scalar matrix $\theta(m)$ allows us to consider $\mathbb{Z}$ as effectively a subring of S. We say that R contains $\mathbb{Z}$ **up to isomorphism**.

Let $\varphi : R \to S$ be a homomorphism, where R and S are rings, with zero elements $0_R, 0_S$, respectively, and let

$$K = \varphi^{-1}(0_S) \ (= \{a \in R : \varphi(a) = 0_S\}). \tag{1.11}$$

We refer to K as the **kernel** of the homomorphism φ, and write it as $\ker \varphi$.

If $a, b \in K$, then $\varphi(a) = \varphi(b) = 0$ and so certainly $\varphi(a - b) = 0$; hence $a - b \in K$. If $r \in R$ and $a \in K$, then $\varphi(ra) = \varphi(r)\varphi(a) = \varphi(r)0 = 0$. (See Exercise 1.1.) Hence $ra \in K$. We deduce that *the kernel of a homomorphism is an ideal.*

In fact the last remark records only one of the ways in which the notions of homomorphism and ideal are linked. Let I be an ideal of a ring R, and let $a \in R$. The set $a + I = \{a + x : x \in I\}$ is called the **residue class of** a **modulo** I. We now show that, for all a, b in R,

$$a + I = b + I \iff a - b \in I, \tag{1.12}$$

and

$$(a + I) + (b + I) = (a + b) + I, \quad (a + I)(b + I) \subseteq ab + I. \tag{1.13}$$

To prove the first of these statements, suppose that $a + I = b + I$. Then, in particular, $a = a + 0 \in a + I = b + I$, and so there exists x in I such that $a = b + x$. Thus $a - b = x \in I$. Conversely, suppose that $a - b \in I$. Then, for all x in I, we have that $a + x = b + y$, where $y = (a - b) + x \in I$. Thus $a + I \subseteq b + I$, and the reverse inclusion is proved in the same way.

To prove the first statement in (1.13), let $x, y \in I$ and let

$$u = (a + x) + (b + y) \in (a + I) + (b + I).$$

Then $u = (a + b) + (x + y) \in (a + b) + I$. Conversely, if $z \in I$ and $v = (a + b) + z \in (a + b) + I$, then $v = (a + z) + (b + 0) \in (a + I) + (b + I)$.

The second statement follows in a similar way. Let $x, y \in I$ and let $u = (a + x)(b + y) \in (a + I)(b + I)$. Then $u = ab + (ay + xb + xy) \in ab + I$.

The set R/I of all residue classes modulo I forms a ring with respect to the operations

$$(a + I) + (b + I) = (a + b) + I, \quad (a + I)(b + I) = ab + I, \tag{1.14}$$

called the **residue class ring** modulo I. The verifications are routine. The zero element is $0 + I = I$; the negative of $a + I$ is $-a + I$. The mapping $\theta_I : R \to R/I$, given by

$$\theta_I(a) = a + I \quad (a \in R), \tag{1.15}$$

is a homomorphism onto R/I, with kernel I. It is called the **natural homomorphism** from R onto R/I.

The motivating example of a residue class ring is the ring $\mathbb{Z}_n$ of integers mod n. Here the ideal is $\langle n \rangle = n\mathbb{Z}$, the set of integers divisible by n, and the elements of $\mathbb{Z}_n$ are the classes $a + \langle n \rangle$, with $a \in \mathbb{Z}$. There are exactly n classes, namely

$$\langle n \rangle, \ 1 + \langle n \rangle, \ 2 + \langle n \rangle, \ \ldots, \ (n - 1) + \langle n \rangle.$$

A strong connection with number theory is revealed by the following theorem:

Theorem 1.8

Let n be a positive integer. The residue class ring $\mathbb{Z}_n = \mathbb{Z}/\langle n \rangle$ is a field if and only if n is prime.

Proof

Suppose first that n is not prime. Then $n = rs$, where $1 < r < n$ and $1 < s < n$. Then $r + \langle n \rangle \neq 0 + \langle n \rangle$ and $s + \langle n \rangle \neq 0 + \langle n \rangle$, but

$$(r + \langle n \rangle)(s + \langle n \rangle) = n + \langle n \rangle = 0 + \langle n \rangle \, .$$

Thus $\mathbb{Z}_n$ contains divisors of 0, and so is certainly not a field.

Now let p be a prime, and suppose that $(r + \langle p \rangle)(s + \langle p \rangle) = 0 + \langle p \rangle$. Then $p \mid rs$, and so (since p is prime) either $p \mid r$ or $p \mid s$. That is, either $r + \langle p \rangle = 0$ or $s + \langle p \rangle = 0$. Thus $\mathbb{Z}_p$ has no divisors of zero, and so is an integral domain. By Exercise 1.5, $\mathbb{Z}_p$ is a field. $\qquad\square$

The next theorem, which has counterparts in many branches of algebra, tells us that every homomorphic image of a ring R is isomorphic to a suitably chosen residue class ring:

Theorem 1.9

Let R be a commutative ring, and let φ be a homomorphism from R onto a commutative ring S, with kernel K. Then there is an isomorphism $\alpha : R/K \to S$ such that the diagram

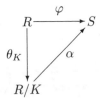

commutes.

Proof

Define α by the rule that

$$\alpha(a + K) = \varphi(a) \quad (a + K \in R/K) \, .$$

The mapping is both well-defined and injective, for

$$a + K = b + K \iff a - b \in K \iff \varphi(a - b) = 0 \iff \varphi(a) = \varphi(b) \, .$$

It clearly maps onto S, since φ is onto. It is a homomorphism, since

$$\alpha\big((a+K)+(b+K)\big) = \alpha\big((a+b)+K\big) = \varphi(a+b)$$
$$= \varphi(a) + \varphi(b) = \alpha(a+K) + \alpha(b+K)\,,$$

and

$$\alpha\big((a+K)(b+K)\big) = \alpha(ab+K) = \varphi(ab) = \varphi(a)\varphi(b) = \alpha(a+K)\alpha(b+K)\,.$$

Hence α is an isomorphism. The commuting of the diagram is clear, since, for all a in R,

$$\alpha\big(\theta_K(a)\big) = \alpha(a+K) = \varphi(a)\,,$$

and so $\alpha \circ \theta_K = \varphi$. □

EXERCISES

1.9. Let a be an element of a ring R. Show that $a + a = a$ implies $a = 0$.

1.10. Show that the definitions (1.1) and (1.2) of a subring are equivalent.

1.11. Show that the definition (1.1) is equivalent to the definition of a subring U of a ring R as a subset of R which is a ring with respect to the operations $+$ and . of R.

1.12. Show that (1.3) is equivalent to the definition of a subfield as a subring which is a field.

1.13. Show that a commutative ring with unity having no proper ideals is a field.

1.14. Show that $\mathbb{Q}(i\sqrt{3}) = \{a + bi\sqrt{3} : a, b \in \mathbb{Q}\}$ is a subfield of $\mathbb{C}$.

1.15. (i) Show that the set

$$K = \left\{ \left(\begin{array}{cc} a & b \\ -3b & a \end{array} \right) : a, b \in \mathbb{Q} \right\}$$

is a field with respect to matrix addition and multiplication.

(ii) Show that K is isomorphic to the field $\mathbb{Q}(i\sqrt{3})$ defined in the previous exercise.

1.16. Show that the set $\mathbb{R}(i\sqrt{3}) = \{a + bi\sqrt{3} : a, b \in \mathbb{R}\}$ is a subfield of $\mathbb{C}$. Is it true that $\mathbb{R}(\sqrt{3}) = \{a + b\sqrt{3} : a, b \in \mathbb{R}\}$ is a subfield of $\mathbb{R}$?

1.17. Let $\varphi : K \to L$ be a non-zero homomorphism, where K and L are fields. Show that φ is a monomorphism.

1.18. Let $\varphi : R \to S$ be a non-zero homomorphism, where R, S are commutative rings with unity, with unity elements 1_R, 1_S, respectively. If R and S are integral domains, show that $\varphi(1_R) = 1_S$. Show by an example that this need not hold if the integral domain condition is dropped.

1.3 The Field of Fractions of an Integral Domain

From Exercise 1.5 we know that every finite integral domain is a field. In this section we show how to construct a field out of an arbitrary integral domain.

Let D be an integral domain. Let

$$P = D \times (D \setminus \{0\}) = \{(a,b) : a, b \in D, \ b \neq 0\}.$$

Define a relation $\equiv$ on the set P by the rule that

$$(a,b) \equiv (a',b') \ \text{ if and only if } \ ab' = a'b.$$

Lemma 1.10

The relation $\equiv$ is an equivalence.

Proof

We must prove (see [13]) that, for all (a,b), (a',b'), (a'',b'') in P,

(i) $(a,b) \equiv (a,b)$ (the **reflexive** law);

(ii) $(a,b) \equiv (a',b') \ \Rightarrow \ (a',b') \equiv (a,b)$ (the **symmetric** law);

(iii) $(a,b) \equiv (a',b')$ and $(a',b') \equiv (a'',b'') \ \Rightarrow \ (a,b) \equiv (a'',b'')$ (the **transitive** law).

The properties (i) and (ii) are immediate from the definition. As for (iii), from $(a,b) \equiv (a',b')$ and $(a',b') \equiv (a'',b'')$ we have that $ab' = a'b$ and $a'b'' = a''b'$. Hence

$$b'(ab'') = (ab')b'' = a'bb'' = b(a'b'') = ba''b' = b'(a''b).$$

Since $b' \neq 0$, we can use the cancellation axiom to obtain $ab'' = a''b$, and so $(a,b) \equiv (a'',b'')$. $\square$

The quotient set $P/\equiv$ is denoted by $Q(D)$. Its elements are equivalence classes $[a,b] = \{(x,y) \in P : (x,y) \equiv (a,b)\}$, and, for reasons that will become obvious, we choose to denote the classes by fraction symbols a/b. Two classes are equal if their (arbitrarily chosen) representative pairs in the set P are equivalent:

$$\frac{a}{b} = \frac{c}{d} \text{ if and only if } ad = bc.$$

In particular, note that

$$\frac{a}{b} = \frac{ka}{kb}$$

for all $k \neq 0$ in D.

We define addition and multiplication in $Q(D)$ by the rules

$$\frac{a}{b} + \frac{c}{d} = \frac{ad+bc}{bd}, \quad \frac{a}{b} \cdot \frac{c}{d} = \frac{ac}{bd}. \tag{1.16}$$

Lemma 1.11

The addition and multiplication defined by (1.16) are well-defined.

Proof

Suppose that $a/b = a'/b'$ and $c/d = c'/d'$. Then $ab' = a'b$ and $cd' = c'd$, and so

$$(ad+bc)b'd' = ab'dd' + bb'cd' = a'bdd' + bb'c'd = (a'd' + b'c')bd.$$

Hence

$$\frac{a}{b} + \frac{c}{d} = \frac{ad+bc}{bd} = \frac{a'd' + b'c'}{b'd'} = \frac{a'}{b'} + \frac{c'}{d'}.$$

Similarly,

$$(ac)(b'd') = (ab')(cd') = (a'b)(c'd) = (a'c')(bd),$$

and so

$$\frac{a}{b} \cdot \frac{c}{d} = \frac{a'}{b'} \cdot \frac{c'}{d'}.$$

$\square$

These operations turn $Q(D)$ into a commutative ring with unity. The verifications are tedious but not difficult. For example,

$$\frac{a}{b}\left(\frac{c}{d} + \frac{e}{f}\right) = \frac{a}{b} \cdot \frac{cf+de}{df} = \frac{acf+ade}{bdf},$$

$$\frac{a}{b} \cdot \frac{c}{d} + \frac{a}{b} \cdot \frac{e}{f} = \frac{ac}{bd} + \frac{ae}{bf} = \frac{acbf + aebd}{b^2 df} = \frac{acf+ade}{bdf}.$$

The zero element is $0/1$ ($= 0/b$ for all $b \neq 0$ in D). The unity element is $1/1$ ($= b/b$ for all $b \neq 0$ in D). The negative of a/b is $(-a)/b$.

The ring $Q(D)$ is in fact a field, since for all a/b with $a \neq 0$ we have that

$$\frac{a}{b} \cdot \frac{b}{a} = \frac{ab}{ab} = \frac{1}{1}.$$

We refer to the field $Q(D)$ as the **field of fractions** of the domain D.

Lemma 1.12

The mapping $\varphi : D \to Q(D)$ given by

$$\varphi(a) = \frac{a}{1} \quad (a \in D) \tag{1.17}$$

is a monomorphism.

Proof

From (1.16) it is clear that

$$\varphi(a) + \varphi(b) = \frac{a}{1} + \frac{b}{1} = \frac{a+b}{1} = \varphi(a+b), \quad \varphi(a)\varphi(b) = \frac{a}{1} \cdot \frac{b}{1} = \frac{ab}{1} = \varphi(ab).$$

Also,

$$\varphi(a) = \varphi(b) \;\Rightarrow\; \frac{a}{1} = \frac{b}{1} \;\Rightarrow\; a = b.$$

$\square$

If we identify $a/1$ with a, we can regard $Q(D)$ as containing D as a subring. The field $Q(D)$ is the **smallest** field containing D, in the following sense:

Theorem 1.13

Let D be an integral domain, let φ be the monomorphism from D into $Q(D)$ given by (1.17) and let K be a field with the property that there is a monomorphism θ from D into K. Then there exists a monomorphism $\psi : Q(D) \to K$ such that the diagram

commutes.

Proof

Define a mapping $\psi : Q(D) \to K$ by the rule that

$$\psi\left(\frac{a}{b}\right) = \frac{\theta(a)}{\theta(b)}.$$

(Note that $\theta(b) \neq 0$, since θ is a monomorphism.) This is well-defined and one-to-one, since

$$\frac{a}{b} = \frac{c}{d} \iff ad = bc \iff \theta(a)\theta(d) = \theta(b)\theta(c) \iff \frac{\theta(a)}{\theta(b)} = \frac{\theta(c)}{\theta(d)},$$

and it is a homomorphism, since

$$\psi\left(\frac{a}{b} + \frac{c}{d}\right) = \psi\left(\frac{ad + bc}{bd}\right) = \frac{\theta(ad + bc)}{\theta(bd)} = \frac{\theta(a)\theta(d) + \theta(b)\theta(c)}{\theta(b)\theta(d)}$$
$$= \frac{\theta(a)}{\theta(b)} + \frac{\theta(c)}{\theta(d)} = \psi\left(\frac{a}{b}\right) + \psi\left(\frac{c}{d}\right),$$

and similarly

$$\psi\left(\frac{a}{b} \cdot \frac{c}{d}\right) = \psi\left(\frac{a}{b}\right)\psi\left(\frac{c}{d}\right).$$

The commuting of the diagram is clear, since, for all a in D,

$$\psi(\varphi(a)) = \psi\left(\frac{a}{1}\right) = \frac{\theta(a)}{\theta(1)} = \theta(a).$$

$\square$

More informally, Theorem 1.3 tells us that any field containing D contains (up to isomorphism) the field $Q(D)$.

When $D = \mathbb{Z}$, it is clear that $Q(D) = \mathbb{Q}$. This is the classical example of the field of quotients, but we shall soon see that it is not the only one.

EXERCISES

1.19. Verify the associativity of addition in $Q(D)$.

1.20. What happens to the construction of $Q(D)$ if D is a field?

1.4 The Characteristic of a Field

In a ring R containing an element a it is reasonable to denote $a + a$ by $2a$, and, more generally, if n is a natural number we may write na for the sum $a + a + \cdots + a$ (n summands). If we define $0a = 0_R$ and $(-n)a$ to be $n(-a)$, we can give a meaning to na for every integer n. The following properties are easy to establish: for $m, n \in \mathbb{Z}$ and $a, b \in R$,

$$(m+n)a = ma + na, \quad m(a+b) = ma + mb, \quad (mn)a = m(na),$$

$$m(ab) = (ma)b = a(mb), \quad (ma)(nb) = (mn)(ab). \tag{1.18}$$

Consider a commutative ring R with unity element 1_R. Then there are two possibilities: either

(i) the elements $m \, 1_R$ ($m = 1, 2, 3, \ldots$) are all distinct; or

(ii) there exist m, n in $\mathbb{N}$ such that $m \, 1_R = (m + n) \, 1_R$.

In the former case we say that R has **characteristic** zero, and write $\operatorname{char} R = 0$. In the latter case we notice that $m \, 1_R = (m + n) \, 1_R = m \, 1_R + n \, 1_R$, and so $n \, 1_R = 0_R$. The least positive n for which this holds is called the **characteristic** of the ring R. Note that, if R is a ring of characteristic n, then $na = 0_R$ for all a in R, for $na = (n \, 1_R)a = 0a = 0$. We write $\operatorname{char} R = n$.

If R is a field, we can say more:

Theorem 1.14

The characteristic of a field is either 0 or a prime number p.

Proof

The former possibility can certainly occur: $\mathbb{Q}$, $\mathbb{R}$ and $\mathbb{C}$ are all fields of characteristic 0. Let K be a field and suppose that $\operatorname{char} K = n \neq 0$, where n is not prime. Then $n = rs$, where $1 < r < n$, $1 < s < n$, and the minimal property of n implies that $r \, 1_K \neq 0_K$, $s \, 1_K \neq 0_K$. On the other hand, from 1.18 we deduce that

$$(r \, 1_K)(s \, 1_K) = (rs) \, 1_K = n \, 1_K = 0_K,$$

and this is impossible, since K, being a field, has no zero divisors. $\qquad\square$

Let K be a field with characteristic 0. Then the elements $n1_K$ ($n \in \mathbb{Z}$) are all distinct, and form a subring of K isomorphic to $\mathbb{Z}$. Indeed, the set

$$P(K) = \{m1_K / n1_f : m, n \in \mathbb{Z}, \ n \neq 0\} \tag{1.19}$$

is a subfield of K isomorphic to $\mathbb{Q}$. Any subfield of K must contain 1 and 0 and so must contain $P(K)$, which is called the **prime subfield** of K.

If K has prime characteristic p, the prime subfield is

$$P(K) = \{1_K, 2(1_K), \ldots, (p-1)(1_K)\}, \tag{1.20}$$

and this is isomorphic to $\mathbb{Z}_p$.

The fields $\mathbb{Q}$ and $\mathbb{Z}_p$ play a central role in the theory of fields. They have no proper subfields, and every field contains as a subfield an isomorphic copy of one or other of them. We frequently want to express this my saying that every field of characteristic 0 is an **extension** of $\mathbb{Q}$, and every field of prime characteristic p is an **extension** of $\mathbb{Z}_p$.

We record these observations formally in a theorem:

Theorem 1.15

Let K be a field. Then K contains a prime subfield $P(K)$ contained in every subfield. If char $K = 0$ then $P(K)$, described by (1.19), is isomorphic to $\mathbb{Q}$. If char $K = p$, a prime number, then $P(K)$, described by (1.20), is isomorphic to $\mathbb{Z}_p$.

Remark 1.16

Given an element a of a field K, we sometimes like to denote $a/(n\,1)$ simply by a/n. If char $K = 0$ this is no problem, but if char $K = p$ then we cannot assign a meaning to a/n if n is a multiple of p. Thus, for example, the formula

$$xy = \frac{1}{4}\left((x+y)^2 - (x-y)^2\right)$$

is not valid in a field of characteristic 2, since the quantity on the right reduces to $0/0$ and so is undefined.

In fields with finite characteristic we encounter some surprising formulae:

Theorem 1.17

Let K be a field of characteristic p. Then, for all x, y in K,

$$(x+y)^p = x^p + y^p.$$

Proof

By the binomial theorem, valid in any commutative ring with unity (see Exercise 1.23), we have that

$$(x + y)^p = \sum_{r=0}^{p} \binom{p}{r} x^{n-r} y^r . \tag{1.21}$$

For $r = 1, \ldots, p - 1$, the binomial coefficient

$$\binom{p}{r} = \frac{p(p-1) \ldots (p-r+1)}{r!}$$

is an integer, and so $r!$ divides $p(p-1) \ldots (p-r+1)$. Since p is prime and $r < p$, no factor of $r!$ can be divisible by p. Hence $r!$ divides $(p-1) \ldots (p-r+1)$, and so $\binom{p}{r}$ is an integer divisible by p. Hence, for $r = 1, \ldots, p - 1$,

$$\binom{p}{r} x^{n-r} y^r = 0 ,$$

and so, in (1.21), only the first and last terms survive.   □

Remark 1.18

The fields $\mathbb{Z}_p = \mathbb{Z}/\langle p \rangle$ are important building blocks in field theory. We usually find it convenient to write $\mathbb{Z}_p = \{0, 1, \ldots, p-1\}$, with addition and multiplication carried out **modulo** p. So, for example, the multiplication table for $\mathbb{Z}_5$ is

	0	1	2	3	4
0	0	0	0	0	0
1	0	1	2	3	4
2	0	2	4	1	3
3	0	3	1	4	2
4	0	4	3	2	1

When it comes to $\mathbb{Z}_3$, it is usually more convenient to write $\mathbb{Z}_3 = \{0, 1, -1\}$. Again, we might at times find it convenient to write $\mathbb{Z}_5$ as $\{0, \pm 1, \pm 2\}$, obtaining the table

	0	1	2	-2	-1
0	0	0	0	0	0
1	0	1	2	-2	-1
2	0	2	-1	1	-2
-2	0	-2	1	-1	2
-1	0	-1	-2	2	1

EXERCISES

1.21. Determine the characteristic of the ring $\mathbb{Z}_6$ of integers mod 6, and show that, in $\mathbb{Z}_6$,
$$a^2 = 0 \;\Rightarrow\; a = 0\,.$$
For which integers n does $\mathbb{Z}_n$ have this property?

1.22. Write down the multiplication table for $\mathbb{Z}_7$, and list the inverses of all the non-zero elements.

1.23. Prove, by induction on n, that the **binomial theorem**,
$$(a+b)^n = \sum_{r=0}^{n} \binom{n}{r} a^{n-r} b^r\,,$$
is valid in a commutative ring R with unity.

1.24. Show that, in a field of finite characteristic p,
$$(x-y)^p = x^p - y^p\,.$$

1.25. Let K be a field of characteristic p. By using Theorem 1.17, deduce, by induction on n, that
$$(x \pm y)^{p^n} = x^{p^n} \pm y^{p^n} \;(x, y \in K,\; n \in \mathbb{N})\,.$$

1.5 A Reminder of Some Group Theory

It is perhaps paradoxical, given the extensive list of axioms that define a field, that a serious study of fields requires a knowledge of more general objects. Rings we have encountered already, though in fact we do not need to explore any further than integral domains. More surprisingly, we need to know some group theory. This does not come into play until well through the book, and you may prefer to skip this section and to return to it when the material is needed. For the most part I will give sketch proofs only: more detail can mostly be found in [13]. As the title suggests, this section is a reminder of the basic ideas and definitions. More specialised bits of group theory, not necessarily covered in a first course in abstract algebra, will be explained when they are needed, and some proofs will be consigned to an appendix.

The axioms for a group were recorded in Section 1.1. It follows from these axioms that the element e in (G2) and the element a^{-1} in (G3) are both unique, and that
$$ae = ea = a\,, \quad aa^{-1} = a^{-1}a = a\,.$$

Also, for all $a, b \in G$,
$$(ab)^{-1} = b^{-1} a^{-1} .$$

The group $(G, .)$ is called a **finite group** if the set G is finite. The cardinality $|G|$ of G is called the **order** of the group.

In the usual way, we write $a^2, a^3, \ldots$ (where $a \in G$) for the products $aa, aaa, \ldots$, and we write a^{-n} to mean $(a^{-1})^n = (a^n)^{-1}$. By a^0 we mean the identity element e. A group G is called **cyclic** if there exists an element a in G such that $G = \{a^n : n \in \mathbb{Z}\}$. If the powers a^n are all distinct, G is the **infinite cyclic group**. Otherwise, there is a least $m > 0$ such that $a^m = e$. The division algorithm then implies, for all n in $\mathbb{Z}$, that there exist integers q and r such that
$$a^n = a^{qm+r} = (a^m)^q a^r = a^r ,$$

and $0 \le r \le m - 1$. Thus $G = \{e, a, a^2, \ldots, a^{m-1}\}$, the **cyclic group of order** m. Both the infinite cyclic group and the cyclic group of order m are abelian.

A non-empty subset U of G is called a **subgroup** of G if, for all $a, b \in G$,
$$a, b \in U \;\Rightarrow\; ab \in U , \quad a \in U \;\Rightarrow\; a^{-1} \in U , \tag{1.22}$$

or, equivalently,
$$a, b \in U \;\Rightarrow\; ab^{-1} \in U . \tag{1.23}$$

Every subgroup contains the identity element e. For each element a in the group G, the set $\{a^n : n \in \mathbb{Z}\}$ is a subgroup, called the **cyclic subgroup generated by** a, and denoted by $\langle a \rangle$. If G is finite, this cannot be the infinite cyclic group, and the order of the cyclic subgroup generated by a is called the **order of the element** a. It is the smallest positive integer n such that $a^n = e$, and is denoted by $o(a)$.

Let U be a subgroup of a group G and let $a \in G$. The subset $Ua = \{ua : u \in U\}$ is called a **left coset** of U. Then $Ua = Ub$ if and only if $ab^{-1} \in U$. Among the left cosets is U itself. The distinct left cosets form a **partition** of G: that is, every element of G belongs to exactly one left coset of U. The mapping $u \mapsto ua$ from U into Ua is easily seen to be both one-one and onto, and so, in a finite group, every left coset has the same number of elements as U. Thus

$$|G| = |U| \times (\text{the number of left cosets}) ,$$

and we have **Lagrange's**[3] **theorem**:

Theorem 1.19

If U is a subgroup of a finite group G, then $|U|$ divides $|G|$.

[3] Joseph-Louis Lagrange, 1736–1813.

It follows immediately that, for all a in G, the order of a divides the order of G.

The choice of left cosets above was arbitrary: exactly the same thing can be done with **right cosets** aU. That is not to say that the right coset aU and the left coset Ua are identical, but the *number* of (distinct) right cosets is the same as the number of left cosets; this number is called the **index** of the subgroup.

If $Ua = aU$ for all a, we say that U is a **normal** subgroup of G, and write $a \lhd b$. Equivalently, U is normal, if, for all a in G,

$$a^{-1}Ua = U \,.$$

In this case we can define a group operation on the set of cosets of U:

$$(Ua)(Ub) = U(ab) \,.$$

First, this is a well-defined operation, since, for all u, v in U,

$$(ua)(vb) = u(av)b = u(v'a)b \text{ (for some } v' \text{ in } U, \text{ since } U \text{ is normal)}$$
$$= (uv')(ab) \in U(ab) \,.$$

Associativity is clear, and it is easy to verify that the identity of the group is the coset $U = Ue$, and the inverse of Ua is Ua^{-1}. The group is denoted by G/U, and is called the **quotient group**, or the **factor group**, of G by U.

Let G, H be groups, with identity elements e_G, e_H, respectively. A mapping $\varphi : G \to H$ is called a **homomorphism** if, for all $a, b \in G$

$$\varphi(ab) = \varphi(a)\varphi(b) \,.$$

It is a consequence of this definition that $\varphi(e_G) = e_H$, and that, for all a in G,

$$\varphi(a^{-1}) = \big(\varphi(a)\big)^{-1} \,.$$

If N is a normal subgroup of G, the mapping $\nu_N : G \to G/N$ given by

$$\nu_N(a) = Na \quad (a \in G)$$

is a homomorphism, called the **natural** homomorphism, onto G/N.

If a homomorphism $\varphi : G \to H$ is one-one and onto, we say that it is an **isomorphism**. In such a case $\varphi^{-1} : H \to G$ is also an isomorphism, and we say that H is **isomorphic** to G, writing $H \simeq G$. If φ maps onto H, but is not necessarily one-one, we say that H is a **homomorphic image** of G.

The **kernel** $\ker \varphi$ of φ is defined by

$$\ker \varphi = \varphi^{-1}(e_H) = \{a \in G \,:\, \varphi(a) = e_H\} \,.$$

It is not hard to show that $\ker \varphi$ is a normal subgroup of G. The following theorem (closely analogous to Theorem 1.9) tells us that every homomorphic image of G is isomorphic to a quotient group of G by a suitable normal subgroup:

Theorem 1.20

Let G, H be groups, and let φ be a homomorphism from G onto H, with kernel N. Then there exists a unique isomorphism $\alpha : G/N \to H$ such that the diagram

commutes.

Proof

The mapping $\alpha : Na \mapsto \varphi(a)$ is well-defined, one-one, onto, and a homomorphism – and $\alpha \circ \nu_N = \varphi$. $\square$

EXERCISES

1.26. Show that every subgroup of index 2 is normal.

1.27. Show that, for every $n \geq 2$, the additive group $(\mathbb{Z}_n, +)$ is cyclic.

1.28. Show that every subgroup of a cyclic group is cyclic.

1.29. Consider the group G of order 8 given by the multiplication table

	e	a	b	c	p	q	r	s
e	e	a	b	c	p	q	r	s
a	a	b	c	e	q	r	s	p
b	b	c	e	a	r	s	p	q
c	c	e	a	b	s	p	q	r
p	p	s	r	q	e	c	b	a
q	q	p	s	r	a	e	c	b
r	r	q	p	s	b	a	e	c
s	s	r	q	p	c	b	a	e

(i) Show that $B = \{e, b\}$ and $Q = \{e, q\}$ are subgroups.

(ii) List the left and right cosets of B and of Q, and deduce that B is normal and Q is not.

(iii) Let H be the group given by the table

	e	x	y	z
e	e	x	y	z
x	x	e	z	y
y	y	z	e	x
z	z	y	x	e

Describe a homomorphism φ from G onto H with kernel B.

1.30. Let $g, h \in A$, where A is a finite abelian group. Show that $o(gh)$ divides $o(g)o(h)$. By considering the group given by

	e	a	b	x	y	z
e	e	a	b	x	y	z
a	a	b	e	z	x	y
b	b	e	a	y	z	x
x	x	y	z	e	a	b
y	y	z	x	b	e	a
z	z	x	y	a	b	e

show that this is not necessarily true in a non-abelian group.

1.31. Let G be a group and N a normal subgroup of G. Show that every subgroup H of G/N can be written as K/N, where K is a subgroup of G containing N, and is normal if and only if H is normal.

2

Integral Domains and Polynomials

2.1 Euclidean Domains

In Chapter 3 we shall start our serious study of fields. But first we need to build our toolkit, which involves polynomial rings over fields. These, as we shall see, are integral domains of a particular kind, and it helps to develop some of the abstract theory of these domains before applying the ideas to polynomials.

An integral domain D is called a **euclidean**[1] **domain** if there is a mapping δ from D into the set $\mathbb{N}^0$ of non-negative integers with the property that $\delta(0) = 0$ and, for all a in D and all b in $D \setminus \{0\}$, there exist q, r in D such that

$$a = qb + r \quad \text{and} \quad \delta(r) < \delta(b). \tag{2.1}$$

From the definition it follows that $\delta^{-1}\{0\} = \{0\}$, for if $\delta(b)$ were equal to 0 it would not be possible to find r such that $\delta(r) < \delta(b)$.

The most important example is the ring $\mathbb{Z}$, where $\delta(a)$ is defined as $|a|$, and where the process, known as the **division algorithm**, is the familiar one (which we have indeed already used in Chapter 1) of dividing a by b and obtaining a **quotient** q and a **remainder** r. If b is positive, then there exists q such that

$$qb \leq a < (q+1)b.$$

---

[1] Euclid of Alexandria, c. 325–265 B.C., is best known for his systematisation of geometry, but he also made significant contributions to number theory, including the *euclidean algorithm* described in the text (applied to the positive integers).

Thus $0 \leq a - qb < b$, and so, taking r as $a - qb$, we see that $a = qb + r$ and $|r| < |b|$. If b is negative, then there exists q such that

$$(q + 1)b < a \leq qb.$$

Thus $b < r = a - qb \leq 0$, and so again $a = qb + r$ and $|r| < |b|$. We shall come across another important example later.

An integral domain D is called a **principal ideal domain** if all of its ideals are principal.

Theorem 2.1

Every euclidean domain is a principal ideal domain.

Proof

Let D be a euclidean domain. The ideal $\{0\}$ is certainly principal. Let I be a non-zero ideal, and let b be a non-zero element of I such that

$$\delta(b) = \min \left\{ \delta(x) \, : \, x \in I \setminus 0 \right\}.$$

Let $a \in I$. Then there exist q, r such that $a = qb + r$ and $\delta(r) < \delta(b)$. Since $r = a - qb \in I$, we have a contradiction unless $r = 0$. Thus $a = qb$, and so $I = Db = \langle b \rangle$, a principal ideal. □

Suppose now that a, b are non-zero members of a principal ideal domain D, and let $\langle a, b \rangle = \{sa + tb \, : \, s, t \in D\}$ be the ideal generated by a and b. (See Theorem 1.4.) By our assumption that D is a principal ideal domain, there exists d in D such that $\langle a, b \rangle = \langle d \rangle$. Since $\langle a \rangle \subseteq \langle d \rangle$ and $\langle b \rangle \subseteq \langle d \rangle$, we have, from Theorem 1.5, that $d \mid a$ and $d \mid b$. Since $d \in \langle a, b \rangle$, there exist s, t in D such that $d = sa + tb$. If $d' \mid a$ and $d' \mid b$, then $d' \mid sa + tb$. That is, $d' \mid d$. We say that d is a **greatest common divisor**, or a **highest common factor**, of a and b. It is effectively unique, for, if $\langle a, b \rangle = \langle d \rangle = \langle d^* \rangle$, it follows from Theorem 1.5 (iii) that $d^* \sim d$.

To summarise, d is the greatest common divisor of a and b (write $d = \gcd(a, b)$) if it has the following properties:

(GCD1) $d \mid a$ and $d \mid b$;

(GCD2) if $d' \mid a$ and $d' \mid b$, then $d' \mid d$.

If $\gcd(a, b) \sim 1$, we say that a and b are **coprime**, or **relatively prime**.

In the case of the domain $\mathbb{Z}$, where the group of units is $\{1, -1\}$, we have, for example, that $\langle 12, 18 \rangle = \langle 6 \rangle = \langle -6 \rangle$.

Remark 2.2

A simple modification of the above argument enables us to conclude that, in a principal ideal domain D, every finite set $\{a_1, a_2, \ldots, a_n\}$ has a greatest common divisor.

In the argument leading to the existence of the greatest common divisor, we assert that "there exists d such that $\langle a, b \rangle = \langle d \rangle$," but give no indication of how this element d might be found. If the domain is euclidean, we do have an algorithm.

The Euclidean Algorithm

Suppose that a and b are non-zero elements of a euclidean domain D, and suppose, without loss of generality, that $\delta(b) \leq \delta(a)$. Then there exist $q_1, q_2, \ldots$ and $r_1, r_2, \ldots$ such that

$$\left.\begin{aligned}
a &= q_1 b + r_1, & \delta(r_1) &< \delta(b), \\
b &= q_2 r_1 + r_2, & \delta(r_2) &< \delta(r_1), \\
r_1 &= q_3 r_2 + r_3, & \delta(r_3) &< \delta(r_2), \\
r_2 &= q_4 r_3 + r_4, & \delta(r_4) &< \delta(r_3), \\
&\cdots\cdots
\end{aligned}\right\} \tag{2.2}$$

The process must end with some $r_k = 0$, the final equations being

$$r_{k-3} = q_{k-1} r_{k-2} + r_{k-1}, \quad \delta(r_{k-1}) < \delta(r_{k-2}),$$
$$r_{k-2} = q_k r_{k-1}.$$

Now, from the first equation of (2.2), we deduce that

$$\langle a, b \rangle = \langle b, r_1 \rangle; \tag{2.3}$$

for every element $sa + tb$ in $\langle a, b \rangle$ can be rewritten as $(t + sq_1)b + sr_1 \in \langle b, r_1 \rangle$, and every element $xb + yr_1$ in $\langle b, r_1 \rangle$ can be rewritten as $ya + (x - yq_1)b \in \langle a, b \rangle$. Similarly, the subsequent equations give

$$\langle b, r_1 \rangle = \langle r_1, r_2 \rangle, \ \langle r_1, r_2 \rangle = \langle r_2, r_3 \rangle, \ldots,$$
$$\langle r_{k-3}, r_{k-2} \rangle = \langle r_{k-2}, r_{k-1} \rangle, \ \langle r_{k-2}, r_{k-1} \rangle = \langle r_{k-1} \rangle. \tag{2.4}$$

From (2.3) and (2.4) it follows that $\langle a, b \rangle = \langle r_{k-1} \rangle$, and so r_{k-1} is the (essentially unique) greatest common divisor of a and b.

Example 2.3

Determine the greatest common divisor of 615 and 345, and express it in the form $615x + 345y$.

Solution

$$615 = 1 \times 345 + 270$$
$$345 = 1 \times 270 + 75$$
$$270 = 3 \times 75 + 45$$
$$75 = 1 \times 45 + 30$$
$$45 = 1 \times 30 + 15$$
$$30 = 2 \times 15 + 0 \,.$$

The greatest common divisor is 15, the last non-zero remainder, and

$$15 = 45 - 30 = 45 - (75 - 45) = 2 \times 45 - 75$$
$$= 2 \times (270 - 3 \times 75) - 75 = 2 \times 270 - 7 \times 75$$
$$= 2 \times 270 - 7 \times (345 - 270) = 9 \times 270 - 7 \times 345$$
$$= 9 \times (615 - 345) - 7 \times 345 = 9 \times 615 - 16 \times 345 \,.$$

$\square$

Two elements a and b of a principal ideal domain D are coprime if their greatest common divisor is 1. This happens if and only if there exist s and t in D such that $sa + tb = 1$. For example, 75 and 64 are coprime:

$$75 = 1 \times 64 + 11$$
$$64 = 5 \times 11 + 9$$
$$11 = 1 \times 9 + 2$$
$$9 = 4 \times 2 + 1 \,,$$

and

$$1 = 9 - 4 \times 2 = 9 - 4(11 - 9) = 5 \times 9 - 4 \times 11 = 5(64 - 5 \times 11) - 4 \times 11$$
$$= 5 \times 64 - 29 \times 11 = 5 \times 64 - 29(75 - 64) = 34 \times 64 - 29 \times 75 \,.$$

EXERCISES

2.1. For the following pairs (a, b) of integers, find the greatest common divisor, and express it as $sa + tb$, where $s, t \in \mathbb{Z}$

   (i) $(1218, 846)$;   (ii) $(851, 779)$.

2.2. Show that a commutative ring with unity is embeddable in a field if and only if it is an integral domain.

2.3. For another example of a euclidean domain, consider the set $\Gamma = \{x + yi \,:\, x, y \in \mathbb{Z}\}$ (where $i = \sqrt{-1}$) of **gaussian[2] integers**.

(i) Show that Γ is an integral domain.

(ii) For each $z = x + yi$ in Γ, define $\delta(z) = |x + yi|^2 = x^2 + y^2$. Let $a, b \in \Gamma$, with $b \neq 0$. Then $ab^{-1} = u + iv$, where $u, v \in \mathbb{Q}$. There exist integers u', v' such that $|u - u'| \leq \frac{1}{2}$, $|v - v'| \leq \frac{1}{2}$. Let $q = u' + iv'$. Show that $a = qb + r$, where $r \in \Gamma$ and $\delta(r) \leq \frac{1}{2}\,\delta(b)$.

2.4. Let p be a prime number, and let

$$D_p = \{\tfrac{r}{s} \in \mathbb{Q} \,:\, r, s \text{ are coprime, and } p \nmid s\}.$$

(i) Show that D_p is a subring of $\mathbb{Q}$.

(ii) Describe the units of D_p.

(iii) Show that D_p is a principal ideal domain.

2.2 Unique Factorisation

Let D be an integral domain with group U of units, and let $p \in D$ be such that $p \neq 0$, $p \notin U$. Then p is said to be **irreducible** if it has no proper factors. An equivalent definition in terms of ideals is available, as a result of the following theorem:

Theorem 2.4

Let p be an element of a principal ideal domain D. Then the following statements are equivalent:

(i) p is irreducible;

(ii) $\langle p \rangle$ is a maximal proper ideal of D;

(iii) $D/\langle p \rangle$ is a field.

Proof

(i) $\Rightarrow$ (ii). Suppose that p is irreducible. Then p is not a unit, and so $\langle p \rangle$ is a proper ideal of D. Suppose, for a contradiction, that there is a (principal) ideal

---

[2] Johann Carl Friedrich Gauss, 1777–1855.

$\langle q \rangle$ such that $\langle p \rangle \subset \langle q \rangle \subset D$. Then $p \in \langle q \rangle$, and so $p = aq$ for some non-unit a. This contradicts the supposed irreducibility of p.

(ii) $\Rightarrow$ (iii). Let $a + \langle p \rangle$ be a non-zero element of $D/\langle p \rangle$. Then $a \notin \langle p \rangle$, and so the ideal $\langle a \rangle + \langle p \rangle$ properly contains $\langle p \rangle$. We are assuming that $\langle p \rangle$ is maximal, and so it follows that $\langle a \rangle + \langle p \rangle = \{sa + tp : s, t \in D\} = D$. Hence there exist s, t in D such that $sa + tp = 1$, and from this we deduce that $(s + \langle p \rangle)(a + \langle p \rangle) = 1 + \langle p \rangle$. Thus $D/\langle p \rangle$ is a field.

(iii) $\Rightarrow$ (i). If p is *not* irreducible, then there exist non-units q and r such that $p = qr$. Then $q + \langle p \rangle$ and $r + \langle p \rangle$ are both non-zero elements of $D/\langle p \rangle$, but

$$(q + \langle p \rangle)(r + \langle p \rangle) = p + \langle p \rangle = 0 + \langle p \rangle.$$

Thus $D/\langle p \rangle$ has divisors of zero, and so certainly is not a field.   □

An element d of an integral domain D has a **factorisation into irreducible elements** if there exist irreducible elements $p_1, p_2, \ldots, p_k$ such that $d = p_1 p_2 \ldots p_k$. The factorisation is **essentially unique** if, for irreducible elements $p_1, p_2, \ldots, p_k$ and $q_1, q_2, \ldots, q_l$,

$$d = p_1 p_2 \ldots p_k = q_1 q_2 \ldots q_l$$

implies that $k = l$ and, for some permutation $\sigma : \{1, 2, \ldots, k\} \rightarrow \{1, 2, \ldots, k\}$,

$$p_i \sim q_{\sigma(i)} \quad (i = 1, 2, \ldots, k).$$

An integral domain D is said to be a **factorial domain,** or to be a **unique factorisation domain**, if every non-unit $a \neq 0$ of D has an essentially unique factorisation into irreducible elements. Here again $\mathbb{Z}$, in which the (positive and negative) prime numbers are the irreducible elements, provides a familiar example: $60 = 2 \times 2 \times 3 \times 5$, and the factorisation is essentially unique, for nothing more different than (say) $(-2) \times (-5) \times 3 \times 2$ is possible.

Theorem 2.5

Every principal ideal domain is factorial.

Proof

We begin with a lemma which at first sight deals with something quite different.

Lemma 2.6

In a principal ideal domain there are no infinite ascending chains of ideals.

Proof

In any integral domain D, an ascending chain

$$I_1 \subseteq I_2 \subseteq I_3 \subseteq \cdots$$

of ideals has the property that $I = \bigcup_{j \geq 1} I_j$ is an ideal. To see this, first observe that, if $a, b \in I$, then there exist k, l such that $a \in I_k$, $b \in I_l$, and so $a - b \in I_{\max\{k,l\}} \subseteq I$. Also, if $a \in I$ and $s \in D$, then $a \in I_k$ for some k, and so $sa \in I_k \subseteq I$.

Now suppose that D is a principal ideal domain, and let

$$\langle a_1 \rangle \subseteq \langle a_2 \rangle \subseteq \langle a_3 \rangle \subseteq \cdots \tag{2.5}$$

be an ascending chain of (principal) ideals. From the previous paragraph, we know that the union of all the ideals in this chain must be an ideal, and, by our assumption about D, this must be a principal ideal $\langle a \rangle$. Since $a \in \bigcup_{j \geq 1} \langle a_j \rangle$, we must have that $a \in \langle a_k \rangle$ for some k. Thus $\langle a \rangle \subseteq \langle a_k \rangle$ and, since it is clear that we also have $\langle a_k \rangle \subseteq \langle a \rangle$, it follows that $\langle a \rangle = \langle a_k \rangle$. Hence

$$\langle a_k \rangle = \langle a_{k+1} \rangle = \langle a_{k+2} \rangle \cdots = \langle a \rangle,$$

and so the infinite chain of inclusions (2.5) terminates at $\langle a_k \rangle$. $\qquad\square$

Returning now to the proof of Theorem 2.5, we show first that any $a \neq 0$ in D can be expressed as a product of irreducible elements. Let a be a non-unit in D. Then either a is irreducible, or it has a proper divisor a_1. Similarly, either a_1 is irreducible, or a_1 has a proper divisor a_2. Continuing, we obtain a sequence $a = a_0, a_1, a_2, \ldots$ in which, for $i = 1, 2, \ldots$, a_i is a proper divisor of a_{i-1}. The sequence must terminate at some a_k, since otherwise we would have an infinite ascending sequence

$$\langle a \rangle \subset \langle a_1 \rangle \subset \langle a_2 \rangle \subset \cdots,$$

and Lemma 2.6 would be contradicted. Hence a has a proper irreducible divisor $a_k = z_1$, and $a = z_1 b_1$. If b_1 is irreducible, then the proof is complete. Otherwise we can repeat the argument we used for a to find a proper irreducible divisor z_2 of b_1, and $a = z_1 z_2 b_2$. We continue this process. It too must terminate, since otherwise we would have an infinite ascending sequence

$$\langle a \rangle \subset \langle b_1 \rangle \subset \langle b_2 \rangle \subset \cdots,$$

in contradiction to Lemma 2.6. Hence some b_l must be irreducible, and so $a = z_1 z_2 \ldots z_{l-1} b_l$ is a product of irreducible elements.

To show that the product is essentially unique, we need another lemma:

Lemma 2.7

Let D be a principal ideal domain, let p be an irreducible element in D, and let $a, b \in D$. Then

$$p \mid ab \implies p \mid a \ \text{ or } \ p \mid b.$$

Proof

Suppose that $p \mid ab$ and $p \nmid a$. Then the greatest common divisor of a and p must be 1, and so there exist s, t in D such that $sa + tp = 1$. Hence $sab + tpb = b$, and so, since p clearly divides $sab + tpb$, it follows that $p \mid b$. $\qquad\square$

It is a routine matter to extend this result to products of more than two elements:

Corollary 2.8

Let D be a principal ideal domain, let p be an irreducible element in D, and let $a_1, a_2, \ldots a_m \in D$. Then

$$p \mid a_1 a_2 \ldots a_m \implies p \mid a_1 \ \text{ or } \ p \mid a_2 \ \text{ or } \ \ldots \ \text{ or } \ p \mid a_m.$$

To complete the proof of Theorem 2.5, suppose that

$$p_1 p_2 \ldots p_k \sim q_1 q_2 \ldots q_l, \qquad (2.6)$$

where $p_1, p_2, \ldots, p_k$ and $q_1, q_2, \ldots, q_l$ are irreducible. Suppose first that $k = 1$. Then $l = 1$, since $q_1 q_2 \ldots q_l$ is irreducible, and so $p_1 \sim q_1$. Suppose inductively that, for all $n \geq 2$ and all $k < n$, any statement of the form (2.6) implies that $k = l$ and that, for some permutation σ of $\{1, 2, \ldots, k\}$,

$$q_i \sim p_{\sigma(i)} \ (i = 1, 2, \ldots k).$$

Let $k = n$. Since $p_1 \mid q_1 q_2 \ldots q_l$, it follows from Corollary 2.8 that $p_1 \mid q_j$ for some j in $\{1, 2, \ldots, l\}$. Since q_j is irreducible and p_1 is not a unit, we deduce that $p_1 \sim q_j$, and by cancellation we then have

$$p_2 p_3 \ldots p_n \sim q_1 \ldots q_{j-1} q_{j+1} \ldots q_l.$$

By the induction hypothesis, we have that $n - 1 = l - 1$ and that, for $i \in \{1, 2, \ldots, n\} \setminus \{j\}$, $q_i \sim p_{\sigma(i)}$ for some permutation σ of $\{2, 3, \ldots, n\}$. Hence, extending σ to a permutation σ of $\{1, 2, \ldots, n\}$ by defining $\sigma(1) = j$, we obtain the desired result. $\qquad\square$

As a consequence of Theorem 2.1, we have the following immediate corollary:

Corollary 2.9

Every euclidean domain is factorial.

EXERCISES

2.5. (i) Determine the group of units of Γ, the domain of gaussian integers.

(ii) Express 5 as a product of irreducible elements of Γ.

(iii) Does
$$13 = (2 + 3i)(2 - 3i) = (3 + 2i)(3 - 2i)$$
contradict unique factorisation in Γ?

2.6. Let $R = \{a + bi\sqrt{3} : a, b \in \mathbb{Z}\}$.

(i) Show that R is a subring of $\mathbb{C}$.

(ii) Show that the map $\varphi : R \to \mathbb{Z}$ given by
$$\varphi(a + bi\sqrt{3}) = a^2 + 3b^2$$
preserves multiplication: for all u, v in R,
$$\varphi(uv) = \varphi(u)\varphi(v).$$
Show also that $\varphi(u) > 3$ unless $u \in \{0, 1, -1\}$.

(iii) Show that the units of R are 1 and -1.

(iv) Show that $1 + i\sqrt{3}$ and $1 - i\sqrt{3}$ are irreducible, and deduce that R is not a unique factorisation domain.

2.3 Polynomials

Throughout this section, R is an integral domain and K is a field.

For reasons that will emerge, we begin by describing a polynomial in abstract terms. The more familiar description of a polynomial will appear shortly. A **polynomial** f with coefficients in R is a sequence $(a_0, a_1, \ldots)$, where $a_i \in R$

for all $i \geq 0$, and where only finitely many of $\{a_0, a_1, \ldots\}$ are non-zero. If the last non-zero element in the sequence is a_n, we say that f has **degree** n, and write $\partial f = n$. The entry a_n is called the **leading coefficient** of f. If $a_n = 1$ we say that the polynomial is **monic**. In the case where *all* of the coefficients are 0, it is convenient to ascribe the formal degree of $-\infty$ to the polynomial $(0, 0, 0, \ldots)$, and to make the conventions, for every n in $\mathbb{Z}$,

$$-\infty < n, \quad -\infty + (-\infty) = -\infty, \quad -\infty + n = -\infty. \tag{2.7}$$

Polynomials $(a, 0, 0, \ldots)$ of degree 0 or $-\infty$ are called **constant**. For others of small degree we have names as follows:

∂f	1	2	3	4	5	6
name	linear	quadratic	cubic	quartic	quintic	sextic

(Fortunately we shall have no occasion to refer to "septic" polynomials!) Addition of polynomials is defined as follows:

$$(a_0, a_1, \ldots) + (b_0, b_1, \ldots) = (a_0 + b_0, a_1 + b_1, \ldots).$$

Multiplication is more complicated:

$$(a_0, a_1, \ldots)(b_0, b_1, \ldots) = (c_0, c_1, \ldots),$$

where, for $k = 0, 1, 2, \ldots$,

$$c_k = \sum_{\{(i,j)\,:\,i+j=k\}} a_i b_j.$$

Thus

$$c_0 = a_0 b_0, \quad c_1 = a_0 b_1 + a_1 b_0, \quad c_2 = a_0 b_2 + a_1 b_1 + a_2 b_0, \ldots.$$

With respect to these two operations, the set P of all polynomials with coefficients in R becomes a commutative ring with unity. Most of the ring axioms are easily verified, and it is clear that the zero element is $(0, 0, 0, \ldots)$, the unity element is $(1, 0, 0, \ldots)$ and the negative of $(a_0, a_1, \ldots)$ is $(-a_0, -a_1, \ldots)$. The only axiom that causes significant difficulty is the associativity of multiplication. Let $p = (a_0, a_1, \ldots)$, $q = (b_0, b_1, \ldots)$, $r = (c_0, c_1, \ldots)$ be polynomials. (Recall that, in each case, only finitely many entries are non-zero.) Then $(pq)r = (d_0, d_1, \ldots)$, where, for $m = 0, 1, 2, \ldots$

$$d_m = \sum_{\{(k,l)\,:\,k+l=m\}} \left(\sum_{\{(i,j)\,:\,i+j=k\}} a_i b_j \right) c_l = \sum_{\{(i,j,l)\,:\,i+j+l=m\}} a_i b_j c_l$$

$$= \sum_{\{(i,n)\,:\,i+n=m\}} a_i \left(\sum_{\{(j,l)\,:\,j+l=n\}} b_j c_l \right),$$

which is the mth entry of $p(qr)$. Thus multiplication is associative.

There is a monomorphism $\theta : R \to P$ given by

$$\theta(a) = (a, 0, 0, \ldots) \quad (a \in R).$$

We may identify the constant polynomial $\theta(a) = (a, 0, 0, \ldots)$ with the element a of R.

Let X be the polynomial $(0, 1, 0, 0, \ldots)$. Then the multiplication rule gives $X^2 = (0, 0, 1, 0, \ldots)$, $X^3 = (0, 0, 0, 1, 0, \ldots)$ and, in general,

$$X^n = (x_0, x_1, \ldots), \text{ where } x_m = \begin{cases} 1 & \text{if } m = n \\ 0 & \text{otherwise.} \end{cases}$$

Then a polynomial

$$(a_0, a_1, \ldots, a_n, 0, 0, \ldots)$$

of degree n can be written as

$$\theta(a_0) + \theta(a_1)X + \theta(a_2)X^2 + \cdots + \theta(a_n)X^n,$$

or as

$$a_0 + a_1 X + a_2 X^2 + \cdots + a_n X^n \tag{2.8}$$

if we make the identification of $\theta(a_i)$ with a_i.

We have arrived at the common definition of a polynomial, in which X is regarded as an "indeterminate". The notation (2.8) is certainly useful, and assuredly makes the definition of multiplication seem less arbitrary. It is important, however, to note that we are talking here of *polynomial forms*, wholly determined by the coefficients a_i, and that X is not a member of R, or indeed of anything else, except of course of the ring P of polynomials. We sometimes write $f = f(X)$ and say that it is a **polynomial over R in the indeterminate X**. The ring P of all such polynomials is written $R[X]$. We refer to it simply as the **polynomial ring** of R.

We summarise some of the main facts about polynomials, some of which we already know.

Theorem 2.10

Let D be an integral domain, and let $D[X]$ be the polynomial ring of D. Then

(i) $D[X]$ is an integral domain.

(ii) if $p, q \in D[X]$, then

$$\partial(p + q) \leq \max \{\partial p, \partial q\}.$$

(iii) for all p, q in $D[X]$,

$$\partial(pq) = \partial p + \partial q.$$

(iv) The group of units of $D[X]$ coincides with the group of units of D.

Proof

(i) We have already noted that $D[X]$ is a commutative ring with unity. To show that there are no divisors of 0, suppose that p and q are non-zero polynomials with leading terms a_m, b_n respectively. The product of p and q then has leading term $a_m b_n$. Since D, by assumption, has no zero divisors, the coefficient $a_m b_n$ is non-zero, and so certainly $pq \neq 0$.

(ii) Let p and q be non-zero. Suppose that $\partial p = m$, $\partial q = n$, and suppose, without loss of generality, that $m \geq n$. If $m > n$ then it is clear that the leading term of $p + q$ is a_m, and so $\partial(p + q) = \max\{\partial p, \partial q\}$. If $m = n$, then we may have $a_m + b_m = 0$, and so all we can say is that $\partial(p + q) \leq \max\{\partial p, \partial q\}$. The conventions established in (2.7) ensure that this result holds also if one or both of p, q are equal to 0.

(iii) By the argument in (i), if p and q are non-zero, then $\partial(pq) = m + n = \partial p + \partial q$. If one or both of p and q are zero, then the result holds by the conventions established in (2.7).

(iv) Let $p, q \in D[X]$, and suppose that $pq = 1$. From Part (iii) we deduce that $\partial p = \partial q = 0$. Thus $p, q \in D$, and $pq = 1$ if and only if p and q are in the group of units of D. $\qquad\square$

Since the ring of polynomials over the integral domain D is itself an integral domain, we can repeat the process, and form the ring of polynomials with coefficients in $D[X]$. We need to use a different letter for a new indeterminate, and the new integral domain is $(D[X])[Y]$, more usually denoted by $D[X, Y]$. It consists of polynomials in the two indeterminates X and Y with coefficients in D. This can be repeated, and we obtain the integral domain $D[X_1, X_2, \ldots, X_n]$.

The field of fractions of $D[X]$ consists of **rational forms**

$$\frac{a_0 + a_1 X + \cdots + a_m X^m}{b_0 + b_1 X + \cdots + b_n X^n},$$

where the denominator is not the zero polynomial. The field is denoted by $D(X)$ (with round rather than square brackets). In a similar way one arrives at the field $D(X_1, X_2, \ldots, X_n)$ of rational forms in the n indeterminates $X_1, X_2, \ldots, X_n$, with coefficients in D.

The point already made, that a polynomial is wholly determined by its coefficients, is underlined by the following result:

Theorem 2.11

Let D, D' be integral domains, and let $\varphi : D \to D'$ be an isomorphism. Then the mapping $\hat{\varphi} : D[X] \to D'[X]$ defined by

$$\hat{\varphi}(a_0 + a_1 X + \cdots + a_n X^n) = \varphi(a_0) + \varphi(a_1)X + \cdots + \varphi(a_n)X^n$$

is an isomorphism.

Proof

The proof is routine. □

The isomorphism $\hat{\varphi}$ is called the **canonical extension** of φ. A further extension $\varphi^* : D(X) \to D'(X)$ is defined by

$$\varphi^*(f/g) = \hat{\varphi}(f)/\hat{\varphi}(g) \quad (f/g \in D(X)). \tag{2.9}$$

We shall be especially interested in the ring $K[X]$ of polynomials over a field K. The group of units of $K[X]$ is the group of units of K, namely the group K^* of non-zero elements of the field K, and in the usual way we write $f \sim g$ if $f = ag$ for some a in K^*.

The integral domain $K[X]$ has an important property closely analogous to a property of the domain of integers:

Theorem 2.12

Let K be a field, and let f, g be elements of the polynomial ring $K[X]$, with $g \neq 0$. Then there exist unique elements q, r in $K[X]$ such that $f = qg + r$ and $\partial r < \partial g$.

Proof

If $f = 0$ the result is trivial, since $f = 0g + 0$. So suppose that $f \neq 0$. The proof is by induction on ∂f. First, suppose that $\partial f = 0$, so that $f \in K^*$. If $\partial g = 0$ also, let $q = f/g$ and $r = 0$; otherwise, let $q = 0$ and $r = f$.

Suppose now that $\partial f = n$, and suppose also that the theorem holds for all polynomials f of all degrees up to $n - 1$. If $\partial g > \partial f$, let $q = 0$ and $r = f$. So suppose now that $\partial g \leq \partial f$. Let f, g have leading terms $a_n X^n$, $b_m X^m$, respectively, where $m \leq n$. Then the polynomial

$$h = f - \left(\frac{a_n}{b_m} X^{n-m}\right) g$$

has degree at most $n - 1$, and so we may assume that there exist q_1, r such that $h = q_1 g + r$, with $\partial r < \partial g$. It follows that $f = qg + r$, where $q = q_1 + (a_n/b_m)X^{n-m}$.

To prove uniqueness, suppose that

$$f = qg + r = q'g + r', \text{ with } \partial r, \partial r' < \partial g.$$

Then $r - r' = (q' - q)g$, and so $\partial\big((q' - q)g\big) = \partial(r - r') < \partial g$. By Theorem 2.10, this cannot happen unless $q' - q = 0$. Hence $q = q'$, and consequently $r = r'$ also.   □

Example 2.13

An actual calculation of q and r for a given pair of polynomials f and d involves a procedure reminiscent of a long division sum. Let $f = X^4 - X$ and $d = X^2 + 3X + 2$.

$$
\begin{array}{r}
X^2 - 3X + 7 \\
X^2 + 3X + 2\;\overline{\smash{\big)}\; X^4 \qquad\qquad - X} \\
X^4 + 3X^3 + 2X^2 \\
\hline
-3X^3 - 2X^2 - X \\
-3X^3 - 9X^2 - 6X \\
\hline
7X^2 + 5X \\
7X^2 + 21X + 14 \\
\hline
-16X - 14
\end{array}
$$

Thus $X^4 - X = (X^2 - 3X + 7)(X^2 + 3X + 2) - (16X + 14)$.

Alternatively, one may equate coefficients in the equality

$$X^4 - X = (X^2 + pX + q)(X^2 + 3X + 2) + (rX + s),$$

finding that $p = -3$, $q = 7$, $r = -16$, $s = -14$.

Theorem 2.14

If K is a field, then $K[X]$ is a euclidean domain.

Proof

The map ∂ does not quite have the properties of the map δ involved in the definition of a euclidean domain, but if, for all f in $K[X]$ we define $\delta(f)$ as $2^{\partial f}$, with the convention that $2^{-\infty} = 0$, we have exactly the right properties.   □

As a consequence of Theorem 2.1, Corollary 2.9 and Theorem 2.4 we can summarise the important properties of $K[X]$ as follows:

Theorem 2.15

Let K be a field. Then,

(i) every pair (f, g) of polynomials in $K[X]$ has a greatest common divisor d, which can be expressed as $af + bg$, with a, b in $K[X]$;

(ii) $K[X]$ is a principal ideal domain;

(iii) $K[X]$ is a factorial domain;

(iv) if $f \in K[X]$, then $K[X]/\langle f \rangle$ is a field if and only if f is irreducible.

Example 2.16

The euclidean algorithm is valid in $K[X]$ (if K is a field) but the calculation can be tedious. Taking a very simple case, we consider the polynomials $X^2 + X + 1$ and $X^3 + 2X - 4$ in $\mathbb{Q}[X]$. Then one may calculate that

$$X^3 + 2X - 4 = (X - 1)(X^2 + X + 1) + 2X - 3$$
$$X^2 + X + 1 = \left(\tfrac{1}{2}X + \tfrac{5}{4}\right)(2X - 3) + \tfrac{19}{4},$$

and so the greatest common divisor is $\frac{19}{4}$. Recall, however, that the group of units of $\mathbb{Q}[X]$ is $\mathbb{Q}^* = \mathbb{Q} \setminus \{0\}$, and so $\frac{19}{4} \sim 1$. The two polynomials are coprime. "Unwinding" the algorithm gives

$$\begin{aligned}
\tfrac{19}{4} &= (X^2 + X + 1) - \left(\tfrac{1}{2}X + \tfrac{5}{4}\right)(2X - 3) \\
&= (X^2 + X + 1) - \left(\tfrac{1}{2}X + \tfrac{5}{4}\right)[(X^3 + 2X - 4) - (X - 1)(X^2 + X + 1)] \\
&= \left(\tfrac{1}{2}X^2 + \tfrac{3}{4}X - \tfrac{1}{4}\right)(X^2 + X + 1) - \left(\tfrac{1}{2}X + \tfrac{5}{4}\right)(X^3 + 2X - 4).
\end{aligned}$$

The **irreducible** elements in the ring $K[X]$ of polynomials over K will be a major area of interest in subsequent chapters.

Example 2.17

Since $X^2 + 1$ is irreducible in $\mathbb{R}[X]$, it follows from Theorem 2.15 that $\mathbb{R}[X]/\langle X^2 + 1 \rangle$ is a field. Denote it by K. The elements of K are residue classes of the form $a + bX + \langle X^2 + 1 \rangle$, where $a, b \in \mathbb{R}$. The addition is given simply by the rule

$$\left(a + bX + \langle X^2 + 1 \rangle\right) + \left(c + dX + \langle X^2 + 1 \rangle\right) = (a + c) + (b + d)X + \langle X^2 + 1 \rangle.$$

Multiplication is a little more difficult:

$$
\begin{aligned}
\big(a + bX + \langle X^2 + 1\rangle\big)\big(c + dX + \langle X^2 + 1\rangle\big) \\
= ac + (ad + bc)X + bdX^2 + \langle X^2 + 1\rangle \\
= (ac - bd) + (ad + bc)X + bd(X^2 + 1) + \langle X^2 + 1\rangle \\
= (ac - bd) + (ad + bc)X + \langle X^2 + 1\rangle.
\end{aligned}
$$

This is reminiscent of the rule for adding and multiplying complex numbers. Indeed it is more than reminiscent: the map $\varphi : \mathbb{R}[X]/\langle X^2 + 1\rangle \to \mathbb{C}$, given by

$$
\varphi\big(a + bX + \langle X^2 + 1\rangle\big) = a + bi \quad (a, b \in \mathbb{R}),
$$

is in fact an isomorphism.

We have already emphasised that polynomials, as we have defined them, are *polynomial forms*, entirely determined by their coefficients. For example, if we write $f = a_0 + a_1 X + \cdots + a_n X^n = 0$, we mean that f is the zero polynomial, that is to say, $a_0 = a_1 = \cdots = a_n = 0$. Let D be an integral domain and let $\alpha \in D$. The **homomorphism** σ_α from $D[X]$ into D is defined by

$$
\sigma_\alpha(a_0 + a_1 X + \cdots + a_n X^n) = a_0 + a_1\alpha + \cdots + a_n\alpha^n. \tag{2.10}
$$

The verification that this is a homomorphism is entirely routine, and is omitted. We frequently want to write $\sigma_\alpha(f)$ more simply as $f(\alpha)$.

If $f(\alpha) = 0$, then we say that α is a **root**, or a **zero**, of the polynomial f. The following result is crucial to the understanding of roots and factorisations.

Theorem 2.18 (The Remainder Theorem)

Let K be a field, let $\beta \in K$ and let f be a non-zero polynomial in $K[X]$. Then the remainder upon dividing f by $X - \beta$ is $f(\beta)$. In particular, β is a root of f if and only if $(X - \beta) \mid f$.

Proof

By the division algorithm (Theorem 2.12), there exist q, r in $K[X]$ such that

$$
f = (x - \beta)q + r, \text{ where } \partial r < \partial(x - \beta) = 1. \tag{2.11}
$$

Thus r is a constant. Substituting β for X, we see that $f(\beta) = r$. In particular, $f(\beta) = 0$ if and only if $r = 0$, that is, if and only if $(X - \beta) \mid q$. $\square$

EXERCISES

2.7. Verify the distributive law $f(g+h) = fg + fh$ for a polynomial ring.

2.8. For the following pairs (f, g) of polynomials, find polynomials q, r such that $f = qg + r$, $\quad \partial r < \partial g$.

(i) $f = X^3 + X + 1$, $g = X^2 + X + 1$;

(ii) $f = X^7 + 1$, $g = X^3 + 1$.

2.9. Show that $\mathbb{Z}[X]$ is not a principal ideal domain.

2.10. Show that, even if K is a field, $K[X, Y]$ is not a principal ideal domain.

2.11. For each of the following pairs (f, g) of polynomials, find the greatest common divisor, and express it in the form $pf + qg$, where p and q are polynomials:

(i) $f = X^5 + X^4 - 2X^3 - X^2 + X$, $g = X^3 + X - 2$;

(ii) $f = X^3 + 2X^2 + 7X - 1$, $g = X^2 + 3X + 4$.

2.12. Show that, in $\mathbb{Z}_p[X]$,

$$X(X-1)(X-2)\ldots\big(X - (p-1)\big) = X^p - X.$$

2.13. Let K be an infinite field, and let f, g be polynomials of degree n. Suppose that there exist distinct elements $\alpha_1, \alpha_2, \ldots, \alpha_{n+1}$ in K such that $f(\alpha_i) = g(\alpha_i)$ $(i = 1, 2, \ldots, n + 1)$. Show that $f = g$.

2.4 Irreducible Polynomials

In Example 2.17 we saw a way of constructing the complex field from the real field. This is a very special case of a more general technique.

Theorem 2.19

Let K be a field, and let $g(X)$ be an irreducible polynomial in $K[X]$. Then $K[X]/\langle g(X) \rangle$ is a field containing K up to isomorphism.

Proof

We know from Theorem 2.15 that $K[X]/\langle g(X)\rangle$ is a field. The map $\varphi : K \to K[X]/\langle g(X)\rangle$ given by

$$\varphi(a) = a + \langle g(X)\rangle \quad (a \in K)$$

is easily seen to be a homomorphism. It is even a monomorphism, since

$$a + \langle g(X)\rangle = b + \langle g(X)\rangle \;\Rightarrow\; a - b \in \langle g(X)\rangle \;\Rightarrow\; a = b\,.$$

$\square$

It is clear, therefore, that we will have a highly effective method of construct-ing new fields provided we have a way of identifying irreducible polynomials. Certainly every linear polynomial is irreducible, and if the field of coefficients is the complex field $\mathbb{C}$, that is the end of the matter, for, by the fundamental theorem of algebra (see [8]), every polynomial in $\mathbb{C}[X]$ factorises, essentially uniquely, into linear factors. Linear polynomials, it must be said, are of little interest as far as Theorem 2.19 is concerned, for $K[X]/\langle g(X)\rangle$ coincides with $\varphi(K)$ in this case, and so is isomorphic to K: if $g(X) = X - a$, then, for all f in $K[X]$ we have that $f = q(X - a) + f(a)$, and so $f + \langle g\rangle = f(a) + \langle g\rangle \in \varphi(K)$.

For polynomials in $\mathbb{R}[X]$ the situation is only a little more complicated. Consider a typical polynomial

$$g(X) = a_n X^n + a_{n-1} X^{n-1} + \cdots + a_1 X + a_0 \tag{2.12}$$

in $\mathbb{R}[X]$. If $\gamma \in \mathbb{C} \setminus \mathbb{R}$ is a root, then

$$a_n \gamma^n + a_{n-1} \gamma^{n-1} + \cdots + a_1 \gamma + a_0 = 0\,.$$

Hence the complex conjugate of the left-hand side is zero also. That is, since the coefficients $a_0, a_1, \ldots, a_n$ are real,

$$a_n \bar{\gamma}^n + a_{n-1} \bar{\gamma}^{n-1} + \cdots + a_1 \bar{\gamma} + a_0\,.$$

Thus the non-real roots of the polynomial occur in conjugate pairs, and we obtain a factorisation

$$g(X) = a_n(X - \beta_1)\ldots(X - \beta_r)(X - \gamma_1)(X - \bar{\gamma}_1)\ldots(X - \gamma_s)(X - \bar{\gamma}_s)\,,$$

in $\mathbb{C}[X]$, where $\beta_1, \ldots, \beta_r \in \mathbb{R}$, $\gamma_1, \ldots, \gamma_s \in \mathbb{C} \setminus \mathbb{R}$, $r, s \geq 0$ and $r + 2s = n$. This gives rise to a factorisation

$$a_n(X - \beta_1)\ldots(X - \beta_r)\big(X^2 - (\gamma_1 + \bar{\gamma}_1)X + \gamma_1\bar{\gamma}_1\big)\ldots\big(X^2 - (\gamma_s + \bar{\gamma}_s)X + \gamma_s\bar{\gamma}_s\big)$$

in $\mathbb{R}[X]$. In this factorisation the quadratic factors are irreducible in $\mathbb{R}[X]$, for if they had real linear factors, they would have two distinct factorisations in $\mathbb{C}[X]$, and we know that this cannot happen.

We have proved the following result:

Theorem 2.20

The irreducible elements of the polynomial ring $\mathbb{R}[X]$ are either linear or quadratic. Every polynomial (2.12) in $\mathbb{R}[X]$ has a unique factorisation

$$a_n(X - \beta_1)\ldots(X - \beta_r)(X^2 + \lambda_1 X + \mu_1)\ldots(X^2 + \lambda_s X + \mu_s),$$

in $\mathbb{R}[X]$, where $a_n \in \mathbb{R}$, $r, s \geq 0$ and $r + 2s = n$.

We can of course easily determine whether a quadratic polynomial $aX^2 + bX + c$ in $\mathbb{R}[X]$ is irreducible: it is irreducible if and only if the **discriminant** $b^2 - 4ac < 0$.

This much is relatively straightforward. Unfortunately, we shall be mostly interested in $\mathbb{Q}[X]$, and here the situation is not so easy, for, as we shall see, in $\mathbb{Q}[X]$ there are irreducible polynomials of arbitrarily large degree.

Quadratic polynomials present no great problem:

Theorem 2.21

Let $g(X) = X^2 + a_1 X + a_0$ be a polynomial with coefficients in $\mathbb{Q}$. Then:

(i) if $g(X)$ is irreducible over $\mathbb{R}$, then it is irreducible over $\mathbb{Q}$;

(ii) if $g(X) = (X - \beta_1)(X - \beta_2)$, with $\beta_1, \beta_2 \in \mathbb{R}$, then $g(X)$ is irreducible in $\mathbb{Q}[X]$ if and only if β_1 and β_2 are irrational.

Proof

(i) Let $g(X)$ be irreducible over $\mathbb{R}$. If $g(X) = (X - q_1)(X - q_2)$ were a factorisation in $\mathbb{Q}[X]$, it would also be a factorisation in $\mathbb{R}[X]$, and we would have a contradiction.

(ii) If β_1, β_2 were rational we would have a factorisation in $\mathbb{Q}[X]$, and $g(X)$ would not be irreducible. If β_1, β_2 are irrational, then $(X - \beta_1)(X - \beta_2)$ is the *only* factorisation in $\mathbb{R}[X]$, and so a factorisation in $\mathbb{Q}[X]$ into linear factors is not possible. $\qquad\square$

Remark 2.22

With regard to part (ii) of the theorem, it is clear that, if one or other of β_1, β_2 is irrational, then both are irrational.

Example 2.23

Examine the following polynomials for irreducibility in $\mathbb{R}[X]$ and $\mathbb{Q}[X]$:

$$X^2 + X + 1, \quad X^2 + X - 1, \quad X^2 + X - 2.$$

Solution

The first polynomial is irreducible over $\mathbb{R}$, since the discriminant is -3. It follows that it is irreducible over $\mathbb{Q}$.

The second polynomial factorises over $\mathbb{R}$ as $(X - \beta_1)(X - \beta_2)$, where

$$\beta_1 = \frac{-1 + \sqrt{5}}{2}, \quad \beta_2 = \frac{-1 - \sqrt{5}}{2}.$$

It is irreducible over $\mathbb{Q}$.

The third polynomial factorises over $\mathbb{Q}$ as $(X - 1)(X + 2)$ and so is not irreducible.   □

To take the matter further we need some new ideas. Observe that in Example 2.23 the factorisation of $X^2 + X - 2$ over $\mathbb{Q}$ is in fact a factorisation over $\mathbb{Z}$. This prompts a question.

- Is it possible for a polynomial $p(X)$ in $\mathbb{Z}[X]$ to be irreducible over $\mathbb{Z}$ but not over $\mathbb{Q}$?

The answer is no.

Theorem 2.24 (Gauss's Lemma)

Let f be a polynomial in $\mathbb{Z}[X]$, irreducible over $\mathbb{Z}$. Then f, considered as a polynomial in $\mathbb{Q}[X]$, is irreducible over $\mathbb{Q}$.

Proof

Suppose, for a contradiction, that $f = gh$, with $g, h \in \mathbb{Q}[X]$ and $\partial g, \partial h < \partial f$. Then there exists a positive integer n such that $nf = g'h'$, where $g', h' \in \mathbb{Z}[X]$. Let us suppose that n is the *smallest* positive integer with this property. Let

$$g' = a_0 + a_1 X + \cdots + a_k X^k, \quad h' = b_0 + b_1 X + \cdots + b_l X^l.$$

If $n = 1$, then $g' = g$, $h' = h$, and we have an immediate contradiction. Otherwise, let p be a prime factor of n.

Lemma 2.25

Either p divides all the coefficients of g', or p divides all the coefficients of h'.

Proof

Suppose, for a contradiction, that p does not divide all the coefficients of g', and that p does not divide all the coefficients of h'. Suppose that p divides $a_0, \ldots, a_{i-1}$, but $p \nmid a_i$, and that p divides $b_0, \ldots, b_{j-1}$, but $p \nmid b_j$. The coefficient of X^{i+j} in nf is

$$a_0 b_{i+j} + \cdots + a_i b_j + \cdots a_{i+j} b_0 .$$

In this sum, all the terms preceding $a_i b_j$ are divisible by p, since p divides $a_0, \ldots, a_{j-1}$; and all the terms following $a_i b_j$ are divisible by p, since p divides $b_0, \ldots, b_{j-1}$. Hence only the term $a_i b_j$ is not divisible by p, and it follows that the coefficient of X^{i+j} in nf is not divisible by p. This gives a contradiction, since the coefficients of f are integers, and so certainly all the coefficients of nf are divisible by p.   □

Returning now to the proof of Theorem 2.24, we may suppose, without loss of generality, that $g' = pg''$, where $g'' \in \mathbb{Z}[X]$. It follows that $(n/p)f = g''h'$, and this contradicts the choice of n as the least positive integer with this property. Hence a factorisation over $\mathbb{Q}$ is not possible, and f is irreducible over $\mathbb{Q}$.   □

We have seen that there is no difficulty in determining the irreducibility of quadratic polynomials in $\mathbb{Q}[X]$. Theorem 2.24 makes it reasonably straightforward to deal with monic cubic polynomials over $\mathbb{Z}$.

Example 2.26

Show that $g = X^3 + 2X^2 + 4X - 6$ is irreducible over $\mathbb{Q}$.

Solution

If the polynomial g factorises over $\mathbb{Q}$, then it factorises over $\mathbb{Z}$, and at least one of the factors must be linear:

$$g = X^3 + 2X^2 + 4X - 6 = (X - a)(X^2 + bX + c). \qquad (2.13)$$

Then $ac = 6$ and so $a \in \{\pm 1, \pm 2, \pm 3, \pm 6\}$. If we substitute a for X in g we must have $g(a) = 0$. However, the values of $g(a)$ are as follows:

a	1	-1	2	-2	3	-3	6	-6
$g(a)$	1	-9	14	-10	51	-27	306	-174

Hence the factorisation (2.13) is impossible, and so g is irreducible over $\mathbb{Q}$.   □

This technique will not work for a polynomial of degree exceeding 3, and indeed there is no easy way to determine irreducibility over $\mathbb{Q}$. One important technique, due to Eisenstein[3], is as follows:

Theorem 2.27 (Eisenstein's criterion)

Let
$$f(X) = a_0 + a_1 X + \cdots + a_n X^n$$
be a polynomial in $\mathbb{Z}[X]$. Suppose that there exists a prime number p such that

(i) $p \nmid a_n$,

(ii) $p \mid a_i \quad (i = 0, \ldots, n-1)$,

(iii) $p^2 \nmid a_0$.

Then f is irreducible over $\mathbb{Q}$.

Proof

By Gauss's lemma (Theorem 2.24), it is sufficient to prove that f is irreducible over $\mathbb{Z}$. Suppose, for a contradiction, that $f = gh$, where
$$g = b_0 + b_1 X + \cdots + b_r X^r, \quad h = c_0 + c_1 X + \cdots + c_s X^s,$$
with $r, s < n$ and $r + s = n$. Since $a_0 = b_0 c_0$, it follows from (ii) that $p \mid b_0$ or $p \mid c_0$. Since $p^2 \nmid a_0$, the coefficients b_0 and c_0 cannot both be divisible by p, and we may assume, without loss of generality, that
$$p \mid b_0, \quad p \nmid c_0. \tag{2.14}$$
Suppose inductively that p divides $b_0, b_1, \ldots, b_{k-1}$, where $1 \le k \le r$. Then
$$a_k = b_0 c_k + b_1 c_{k-1} + \cdots + b_{k-1} c_1 + b_k c_0.$$
Since p divides each of $a_k, b_0 c_k, b_1 c_{k-1}, \ldots, b_{k-1} c_1$, it follows that $p \mid b_k c_0$, and hence, from (2.14), $p \mid b_k$.

We conclude that $p \mid b_r$, and so, since $a_n = b_r c_s$, we have that $p \mid a_n$, a contradiction to the assumption (i). Hence f is irreducible. $\qquad\square$

Remark 2.28

It is clear from Theorem 2.27 that there exist irreducible polynomials in $\mathbb{Q}[X]$ of arbitrarily high degree.

---

[3] Ferdinand Gotthold Max Eisenstein, 1823–1852.

Example 2.29

The polynomial $X^5 + 2X^3 + \frac{8}{7}X^2 - \frac{4}{7}X + \frac{2}{7}$ is irreducible over $\mathbb{Q}$, since $7X^5 + 14X^3 + 8X^2 - 4X + 2$ satisfies Eisenstein's criterion, with $p = 2$.

It is sometimes possible to apply the Eisenstein test after a suitable adjustment:

Example 2.30

Show that

$$f(X) = 2X^5 - 4X^4 + 8X^3 + 14X^2 + 7$$

is irreducible over $\mathbb{Q}$.

Solution

The polynomial f does not satisfy the required conditions. If, however, there exists a factorisation $f = gh$ with (say) $\partial g = 3$ and $\partial h = 2$, then

$$7X^5 + 14X^3 + 8X^2 - 4X + 2 = X^5 f(1/X) = \left(X^3 g(1/X)\right)\left(X^2 h(1/X)\right)$$

is a factorisation of $7X^5 + 14X^3 + 8X^2 - 4X + 2$, and from Example 2.29 we know that this cannot happen. $\square$

The next example will eventually prove important:

Example 2.31

Show that, if $p > 2$ is prime, then

$$f(X) = 1 + X + X^2 + \cdots + X^{p-1}$$

is irreducible over $\mathbb{Q}$.

Solution

Observe that $f(X) = (X^p - 1)/(X - 1)$. If $g(X)$ is defined as $f(X + 1)$, it follows that

$$g(X) = \frac{1}{X}\left((X + 1)^p - 1\right) = \sum_{r=0}^{p-1} \binom{p}{r} X^{p-r-1} .$$

As was observed in the proof of Theorem 1.17, the coefficients

$$\binom{p}{1}, \binom{p}{2}, \ldots, \binom{p}{p-1}$$

are all divisible by p. Hence g is irreducible, by the Eisenstein criterion.

If $f = uv$, with $\partial u, \partial v < \partial f$ and $\partial u + \partial v = \partial f$, then

$$g(X) = u(X+1)v(X+1).$$

The factors $u(X+1)$ and $v(X+1)$ are polynomials in X, of the same degrees (respectively) as u and v. We thus have a contradiction, since g is irreducible.

$\square$

Another device for determining irreducibility over $\mathbb{Z}$ (and consequently over $\mathbb{Q}$) is to map the polynomial onto $\mathbb{Z}_p[X]$ for some suitably chosen prime p. Let $g = a_0 + a_1 X + \cdots + a_n X^n \in \mathbb{Z}[X]$, and let p be a prime not dividing a_n. For each i in $\{0, 1, \ldots, n\}$, let $\bar{a}_i$ denote the residue class $a_i + \langle p \rangle$ in the field $\mathbb{Z}_p = \mathbb{Z}/\langle p \rangle$, and write the polynomial $\bar{a}_0 + \bar{a}_1 X + \cdots + \bar{a}_n X^n$ as $\bar{g}$. Our choice of p ensures that $\partial \bar{g} = n$. Suppose that $g = uv$, with $\partial u, \partial v < \partial f$ and $\partial u + \partial v = \partial g$. Then $\bar{g} = \bar{u}\,\bar{v}$. If we can show that $\bar{g}$ is irreducible in $\mathbb{Z}_p[X]$, then we have a contradiction, and we deduce that g is irreducible. The advantage of transferring the problem from $\mathbb{Z}[X]$ to $\mathbb{Z}_p[X]$ is that $\mathbb{Z}_p$ is finite, and the verification of irreducibility is a matter of checking a finite number of cases.

Example 2.32

Show that

$$g = 7X^4 + 10X^3 - 2X^2 + 4X - 5$$

is irreducible over $\mathbb{Q}$.

Solution

If we choose $p = 3$, then, in the notation of the paragraph preceding this example,

$$\bar{g} = X^4 + X^3 + X^2 + X + 1.$$

The elements of $\mathbb{Z}_3$ may be taken as $0, 1, -1$, with $1 + 1 = -1$.

We show first that $\bar{g}$ has no linear factor, for

$$\bar{g}(0) = 1, \quad \bar{g}(1) = -1, \quad \bar{g}(-1) = 1.$$

There remains the possibility that (in $\mathbb{Z}_3[X]$)

$$X^4 + X^3 + X^2 + X + 1 = (X^2 + aX + b)(X^2 + cX + d).$$

Equating coefficients gives

$$
\begin{aligned}
a + c &= 1, & b + ac + d &= 1, \\
bd &= 1, & ad + bc &= 1.
\end{aligned}
$$

Hence either (i) $b = d = 1$ or (ii) $b = d = -1$. In case (i) we deduce that $ac = -1$, and so $a = \pm 1$, $c = \mp 1$. In either case $a + c = 0$, and we have a contradiction. In case (ii) we deduce that $ac = 0$. If $a = 0$ then $c = 1$, and so $1 = ad + bc = b$, a contradiction. Similarly, if $c = 0$ then $a = 1$, and then $1 = ad + bc = d$, again a contradiction.

We have shown that $\bar{g}$ is irreducible over $\mathbb{Z}_3$, and it follows that g is irreducible over $\mathbb{Q}$. $\qquad\square$

Remark 2.33

The choice of the prime p is, of course, crucial. If, in the above example, we had used $p = 2$, we would have obtained $\bar{g} = X^4 + 1$, and in $\mathbb{Z}_2[X]$ this is far from irreducible, since $X^4 + 1 = (X + 1)^4$. It is important to realise that if our $\bar{g}$ is not irreducible then we can draw no conclusion at all.

EXERCISES

2.14. Show that $X^3 + 2X^2 - 3X + 5$ is irreducible over $\mathbb{Q}$.

2.15. Show that

$$X^3 + 3X + 12, \quad X^4 + 2X - 6, \quad X^5 + 5X^2 - 10$$

are irreducible over $\mathbb{Q}$.

2.16. By making suitable transformations, use the Eisenstein criterion to show that

$$5X^4 - 10X^3 + 10X - 3, \quad X^4 + 4X^3 + 3X^2 - 2X + 4$$

are irreducible.

2.17. By using the technique of Example 2.32, show that

$$4X^4 - 2X^2 + X - 5, \quad 3X^4 - 7X + 5$$

are irreducible.

3

Field Extensions

3.1 The Degree of an Extension

In this section it is necessary to have some knowledge of the basic concepts of linear algebra, including linear independence, spanning sets, bases and dimension. See, for example, [3].

If K, L are fields and $\varphi : K \to L$ is a monomorphism, we say that L is an **extension** of K, and we sometimes find it useful to write "$L : K$ is a (field) extension". As we have seen, this is not essentially different from saying that K is a subfield of L, since we may always identify K with its image $\varphi(K)$. Then L can be regarded as a vector space over K, since the vector space axioms

(V1) $(x + y) + z = x + (y + z)$ $(x, y, z \in L)$,

(V2) $x + y = y + x$ $(x, y \in L)$,

(V3) there exists 0 in L such that $x + 0 = x$ $(x \in L)$,

(V4) for all x in L there exists $-x$ in L such that $x + (-x) = 0$,

(V5) $a(x + y) = ax + ay$ $(a \in K,\ x, y \in L)$,

(V6) $(a + b)x = ax + bx$ $(a, b \in K,\ x \in L)$,

(V7) $(ab)x = a(bx)$ $(a, b \in K,\ x \in L)$,

(V8) $1x = x$ $(x \in L)$,

are all consequences of the field axioms for L. Hence there exists a **basis** of L over K. Different bases have the same cardinality, and there is a well-defined

dimension of L, equal to the cardinality of an arbitrarily chosen basis. The term used in field theory for this dimension is the **degree of L over K**, or the **degree of the extension** $L : K$; and we denote it by $[L : K]$. We say that L is a **finite extension** of K if $[L : K]$ is finite; otherwise L is an **infinite extension**.

Example 3.1

The field $\mathbb{R}$ of real numbers is an infinite extension of $\mathbb{Q}$, since any finite extension of $\mathbb{Q}$ is countable, and $\mathbb{R}$ is not. (See [6] for information on infinite cardinal numbers.) We shall return to this issue later in the chapter when transcendental numbers make their appearance. By contrast, the field $\mathbb{C}$ of complex numbers is a finite extension of $\mathbb{R}$, with basis $\{1, i\}$, since every complex number has a unique expression as $a1 + bi$, with $a, b \in \mathbb{R}$. The basis is, of course, not unique: for example, we can write $a + bi$ as $\frac{1}{2}(a + b)(1 + i) + \frac{1}{2}(a - b)(1 - i)$, and so $\{1 + i, 1 - i\}$ is a basis. However, every basis has exactly two elements, and $[\mathbb{C} : \mathbb{R}] = 2$.

Theorem 3.2

Let $L : K$ be a field extension. Then $L = K$ if and only if $[L : K] = 1$.

Proof

This is a standard property of finite-dimensional vector spaces, but for completeness we prove it here.

Suppose first that $L = K$. Then $\{1\}$ is a basis for L over K, since every element x of L is expressible as $x1$, with x in K. Thus $[L : K] = 1$.

Conversely, suppose that $[L : K] = 1$, and that $\{x\}$, where $x \neq 0$, is a basis of L over K. Thus, in particular, there exists a in K such that $1 = ax$, and so $x = 1/a \in K$. For every y in L there exists b in K such that $y = bx = b/a$. Thus $y \in K$. We have shown that $L = K$.   □

Suppose now that we have field extensions $L : K$ and $M : L$. That is, there are monomorphisms $\alpha : K \to L$, $\beta : L \to M$. Then $\beta \circ \alpha : K \to M$ is a monomorphism, and so M is an extension of K. With these definitions we now have the following theorem, in which the equality is intended to include the information that if either of $[M : L]$ and $[L : K]$ is infinite then so is $[M : K]$. We shall make the usual identifications, regarding K as a subfield of L and L as a subfield of M.

Theorem 3.3

Let $L : K$ and $M : L$ be field extensions. Then

$$[M : L]\,[L : K] = [M : K]\,.$$

Proof

Let $\{a_1, a_2, \ldots, a_r\}$ be a linearly independent subset of M over L, and let $\{b_1, b_2, \ldots, b_s\}$ be a linearly independent subset of L over K. We show that

$$\{a_i b_j \ : \ i = 1, 2, \ldots r, \ j = 1, 2, \ldots s\} \tag{3.1}$$

is a linearly independent subset of M over K. For let us suppose that

$$\sum_{i=1}^{r} \sum_{j=1}^{s} \lambda_{ij} a_i b_j = 0\,,$$

with $\lambda_{ij} \in K$ for all i and j. Rewriting this as

$$\sum_{i=1}^{r} \left(\sum_{j=1}^{s} \lambda_{ij} b_j \right) a_i = 0\,,$$

we deduce, since the a_i are linearly independent over L, that

$$\sum_{j=1}^{s} \lambda_{ij} b_j = 0 \quad (i = 1, 2, \ldots, r)\,.$$

Then, since the b_j are linearly independent over K, we conclude that $\lambda_{ij} = 0$ for all i and j.

If either of $[M : L]$ and $[L : K]$ is infinite, then either r or s can be made arbitrarily large, and so the set (3.1) can be made arbitrarily large. Hence $[M : K]$ is infinite. So now suppose that

$$[M : L] = r < \infty\,, \quad [L : K] = s < \infty\,,$$

that $\{a_1, a_2, \ldots, a_r\}$ is a basis of M over L, and that $\{b_1, b_2, \ldots, b_s\}$ is a basis of L over K. For each z in M there exist $\lambda_1, \lambda_2, \ldots, \lambda_r$ in L such that $z = \sum_{i=1}^{r} \lambda_i a_i$. Also, for each λ_i there exist $\mu_{i1}, \mu_{i2}, \ldots, \mu_{is}$ in K such that $\lambda_i = \sum_{j=1}^{s} \mu_{ij} b_j$. Hence

$$z = \sum_{i=1}^{r} \sum_{j=1}^{s} \mu_{ij} (a_i b_j)\,.$$

The set (3.1), being both linearly independent and a spanning set for M over K, is a basis, and so

$$[M : K] = rs = [M : L]\,[L : K]\,.$$

$\square$

The following easy consequence is worth recording at this stage.

Corollary 3.4

Let $K_1, K_2, \ldots, K_n$ be fields, and suppose that $K_{i+1} : K_i$ is an extension, for $1 \le i \le n - 1$. Then

$$[K_n : K_1] = [K_n : K_{n-1}][K_{n-1} : K_{n-2}] \ldots [K_2 : K_1].$$

EXERCISES

3.1. Let $L : K$ and $M : L$ be field extensions, and let $[M : K]$ be finite. Show that

(i) if $[M : K] = [L : K]$, then $M = L$;

(ii) if $[M : L] = [M : K]$, then $L = K$.

3.2. Let $L : K$ be a field extension such that $[L : K]$ is a prime number. Show that there is no subfield E of L such that $K \subset E \subset L$.

3.2 Extensions and Polynomials

We are familiar with the observation that the equation $X^2 = 2$ cannot be solved within the rational field, but has the solutions $\pm\sqrt{2}$ in the field $\mathbb{R}$ of real numbers. In fact its solutions lie within a much smaller field than $\mathbb{R}$, namely, the extension

$$\mathbb{Q}[\sqrt{2}] = \{a + b\sqrt{2} : a, b \in \mathbb{Q}\}$$

of $\mathbb{Q}$. It is not perhaps quite obvious that this is a field, but it is easy to verify the subfield conditions (1.3). If $a + b\sqrt{2}, c + d\sqrt{2} \in \mathbb{Q}[\sqrt{2}]$, then

$$(a + b\sqrt{2}) - (c + d\sqrt{2}) = (a - c) + (b - d)\sqrt{2} \in \mathbb{Q}[\sqrt{2}],$$

and (if $c + d\sqrt{2} \neq 0$)

$$(a + b\sqrt{2})(c + d\sqrt{2})^{-1} = \frac{(a + b\sqrt{2})(c - d\sqrt{2})}{(c + d\sqrt{2})(c - d\sqrt{2})} = u + v\sqrt{2},$$

where

$$u = \frac{ac - 2bd}{c^2 - 2d^2}, \quad v = \frac{bc - ad}{c^2 - 2d^2}.$$

Note that from the irrationality of $\sqrt{2}$ it follows that $c^2 - 2d^2 = 0$ if and only if $c = d = 0$.

This is a special case of a general result, which we now proceed to investigate.

We begin with something quite general. Let K be a subfield of a field L, and let S be a subset of L. Let $K(S)$ be the intersection of all the subfields of L containing $K \cup S$. (There is at least one such subfield, namely L itself.) It is clear that $K(S)$ is the smallest subfield containing $K \cup S$, and we call it the **subfield of L generated over K by** S. If $S = \{\alpha_1, \alpha_2, \ldots, \alpha_n\}$ is finite, we write $K(S)$ as $K(\alpha_1, \alpha_2, \ldots, \alpha_n)$.

Theorem 3.5

The subfield $K(S)$ of the field L coincides with the set E of all elements of L that can be expressed as quotients of finite linear combinations (with coefficients in K) of finite products of elements of S.

Proof

Denote by P the set of all finite linear combinations of finite products of elements of S. If $p, q \in P$, then $p \pm q, pq \in P$. Hence, if $x = p/q$ and $y = r/s$ are typical elements of E, with p, q, r, s in P and $q, s \neq 0$, we see that $x - y = (ps - qr)/(qs) \in E$, and (provided $y \neq 0$) $x/y = (ps)/(qr) \in E$. From (1.3) we deduce that E is a subfield of L containing K and S, and so $K(S) \subseteq E$. Also, any subfield containing K and S must contain all finite products of elements in S, all linear combinations of such products, and all quotients of such linear combinations. In short, it must contain E. Hence, in particular, $K(S) \supseteq E$. $\qquad\square$

Of particular interest is the case where S has just one element α ($\notin K$). Then, from Theorem 3.5, we deduce that $K(\alpha)$ is the set of all quotients of polynomials in α with coefficients in K. We say that $K(\alpha)$ is a **simple extension** of K. The link with polynomials is important, as the next result shows:

Theorem 3.6

Let L be a field, let K be a subfield and let $\alpha \in L$. Then either

(i) $K(\alpha)$ is isomorphic to $K(X)$, the field of all rational forms with coefficients in K; or

(ii) there exists a unique monic irreducible polynomial m in $K[X]$ with the property that, for all f in $K[X]$,

   (a) $f(\alpha) = 0$ if and only if $m \mid f$;

   (b) the field $K(\alpha)$ coincides with $K[\alpha]$, the ring of all polynomials in α with coefficients in K; and

   (c) $[K[\alpha] : K] = \partial m$.

Proof

Suppose first that there is no non-zero polynomial f in $K[X]$ such that $f(\alpha) = 0$. (This means in particular that $\alpha \notin K$, since in that case we may take f as $X - \alpha$.) Then there is a mapping $\varphi : K(X) \to K(\alpha)$ given by

$$\varphi(f/g) = f(\alpha)/g(\alpha) \,,$$

(for we are assuming that $g(\alpha) = 0$ only if g is the zero polynomial). It is routine to verify that φ is a homomorphism, and it clearly maps onto $K(\alpha)$. To see that it is well defined and one-to-one, suppose that f, g, p, q are polynomials, with $g, q \neq 0$. Then

$$\begin{aligned}
\varphi(f/g) = \varphi(p/q) &\iff f(\alpha)q(\alpha) - p(\alpha)g(\alpha) = 0 \ \text{in} \ L \\
&\iff fq - pg = 0 \ \text{in} \ K[X] \\
&\iff f/g = p/q \ \text{in} \ K(X) \,.
\end{aligned}$$

Now suppose that there does exist a non-zero polynomial g such that $g(\alpha) = 0$. Indeed, let us suppose that g is a polynomial *with least degree* having this property. If a is the leading coefficient of g, then g/a is a monic polynomial. Denote g/a by m. Certainly $m(\alpha) = 0$.

It is clear that $f(\alpha) = 0$ if $m \mid f$. Conversely, suppose that $f(\alpha) = 0$. Then, by Theorem 2.12, $f = qm + r$, where $\partial r < \partial m$. Now

$$0 = f(\alpha) = q(\alpha)m(\alpha) + r(\alpha) = 0 + r(\alpha) = r(\alpha) \,.$$

Since $\partial r < \partial m$, this gives a contradiction unless r is the zero polynomial. Hence $f = qm$, and so $m \mid f$.

To show that m is unique, suppose that m' is another polynomial with the same properties. Then $m(\alpha) = m'(\alpha) = 0$ and so $m \mid m'$ and $m' \mid m$. Since both polynomials are monic, we conclude that $m' = m$.

To show that m is irreducible, suppose, for a contradiction, that there exist polynomials p and q such that $pq = m$, with $\partial p, \partial q < \partial m$. Then $p(\alpha)q(\alpha) = m(\alpha) = 0$, and so either $p(\alpha) = 0$ or $q(\alpha) = 0$. This is impossible, since both p and q are of smaller degree than m.

Next, consider a typical element $f(\alpha)/g(\alpha)$ in $K(\alpha)$, where $g(\alpha) \neq 0$. Then m does not divide g, and it follows, since m has no divisors other than itself and 1, that the greatest common divisor of g and m is 1. Hence, by Theorem 2.15, there exist polynomials a, b such that $ag + bm = 1$, and so, substituting α for X, we have $a(\alpha)g(\alpha) = 1$. Thus

$$\frac{f(\alpha)}{g(\alpha)} = f(\alpha)a(\alpha) \in K[\alpha].$$

Finally, suppose that $\partial m = n$, and let $p(\alpha) \in K[\alpha] = K(\alpha)$, where p is a polynomial. Then $p = qm + r$, where $\partial r < \partial m = n$. It follows that $p(\alpha) = r(\alpha)$, and so there exist $c_0, c_1, \ldots c_{n-1}$ (the coefficients of r, some of which may, of course, be zero) in K such that $p(\alpha) = c_0 + c_1\alpha + \cdots + c_{n-1}\alpha^{n-1}$. Hence $\{1, \alpha, \ldots, \alpha^{n-1}\}$ is a spanning set for $K[\alpha]$.

Moreover, the set $\{1, \alpha, \ldots, \alpha^{n-1}\}$ is linearly independent over K, for if elements $a_0, a_1, \ldots, a_{n-1}$ of K are such that $a_0 + a_1\alpha + \cdots + a_{n-1}\alpha^{n-1} = 0$, then $a_0 = a_1 = \cdots = a_{n-1} = 0$, since otherwise we would have a non-zero polynomial $p = a_0 + a_1X + \cdots + a_{n-1}X^{n-1}$ of degree at most $n - 1$ such that $p(\alpha) = 0$. Thus $\{1, \alpha, \ldots, \alpha^{n-1}\}$ is a basis of $K(\alpha)$ over K, and so $[K(\alpha) : K] = n$. $\qquad\square$

The polynomial m defined above is called the **minimum polynomial** of the element α.

Remark 3.7

If m' is another monic polynomial of degree n such that $m'(\alpha) = 0$, then $m \mid m'$ implies that $m = m'$. Thus, if we know that $[K[\alpha] : K] = n$ and if we find a monic polynomial g of degree n such that $g(\alpha) = 0$, then g must be the minimum polynomial of α.

From the proof of Theorem 3.6 we see that every $f(\alpha)/g(\alpha)$ in $K(\alpha)$ is expressible as a linear combination of $1, \alpha, \ldots, \alpha^{n-1}$, with coefficients in K. To find this expression for a given element of $K(\alpha)$, we can follow the procedure in the proof of the theorem, but there is usually a simpler way.

Example 3.8

Let α be an element of $\mathbb{C}$ with minimum polynomial $X^2 + X + 1$ over $\mathbb{Q}$. Show that $\alpha^2 - 1 \neq 0$, and express the element $(\alpha^2 + 1)/(\alpha^2 - 1)$ of $\mathbb{Q}(\alpha)$ in the form $a + b\alpha$, where $a, b \in \mathbb{Q}$.

Solution

Since $\alpha^2 + \alpha + 1 = 0$, we immediately have that $\alpha^2 - 1 = -\alpha - 2 \neq 0$. Hence

$$\frac{\alpha^2 + 1}{\alpha^2 - 1} = \frac{-\alpha}{-\alpha - 2} = \frac{\alpha}{\alpha + 2} = 1 - \frac{2}{\alpha + 2}\,.$$

Dividing $X^2 + X + 1$ by $X + 2$ gives

$$X^2 + X + 1 = (X + 2)(X - 1) + 3\,,$$

and so $(\alpha + 2)(\alpha - 1) = -3$. Hence

$$\frac{1}{\alpha + 2} = -\frac{1}{3}(\alpha - 1)\,,$$

and so

$$\frac{\alpha^2 + 1}{\alpha^2 - 1} = 1 + \frac{2}{3}(\alpha - 1) = \frac{1}{3}(1 + 2\alpha)\,.$$

$\square$

Example 3.9

If K is the field $\mathbb{Q}$ and L the field $\mathbb{C}$, the minimum polynomial of $i\sqrt{3}$ is $X^2 + 3$. Then

$$\mathbb{Q}[i\sqrt{3}] = \{a + bi\sqrt{3} : a, b \in \mathbb{Q}\}\,.$$

The multiplicative inverse of a non-zero element $a + bi\sqrt{3}$ is $a' + b'i\sqrt{3}$, where

$$a' = \frac{a}{a^2 + 3b^2}\,, \qquad b' = \frac{-b}{a^2 + 3b^2}\,.$$

Example 3.10

It might seem that the subfield $\mathbb{Q}(\sqrt{2}, \sqrt{3})$ is not a simple extension, but in fact it coincides with the visibly simple extension $\mathbb{Q}(\sqrt{2} + \sqrt{3})$. It is clear that $\sqrt{2} + \sqrt{3} \in \mathbb{Q}(\sqrt{2}, \sqrt{3})$, and so $\mathbb{Q}(\sqrt{2} + \sqrt{3}) \subseteq \mathbb{Q}(\sqrt{2}, \sqrt{3})$. Conversely, since $(\sqrt{3} + \sqrt{2})(\sqrt{3} - \sqrt{2}) = 1$, it follows that $\sqrt{3} - \sqrt{2} = (\sqrt{3} + \sqrt{2})^{-1} \in \mathbb{Q}(\sqrt{2} + \sqrt{3})$, and it then follows easily that $\sqrt{2}, \sqrt{3} \in \mathbb{Q}(\sqrt{2} + \sqrt{3})$. Hence $\mathbb{Q}(\sqrt{2}, \sqrt{3}) \subseteq \mathbb{Q}(\sqrt{2} + \sqrt{3})$.

We can write $\mathbb{Q}(\sqrt{2}, \sqrt{3})$ as $(\mathbb{Q}[\sqrt{2}])[\sqrt{3}]$. The set $\{1, \sqrt{2}\}$ is clearly a basis for $\mathbb{Q}[\sqrt{2}]$ over $\mathbb{Q}$. Since $\sqrt{3} \notin \mathbb{Q}[\sqrt{2}]$ (see Exercise 2.4), we must have $[\mathbb{Q}(\sqrt{2}, \sqrt{3}) : \mathbb{Q}[\sqrt{2}]] \geq 2$. On the other hand, from the trivial observation that $(\sqrt{3})^2 - 3 = 0$, we conclude that $X^2 - 3$ is the minimum polynomial of $\sqrt{3}$ over $\mathbb{Q}[\sqrt{2}]$ and that $\{1, \sqrt{3}\}$ is a basis. Then, from Theorem 3.3, we deduce that $\{1, \sqrt{2}, \sqrt{3}, \sqrt{6}\}$ is a basis for $\mathbb{Q}(\sqrt{2}, \sqrt{3})$ over $\mathbb{Q}$.

The minimum polynomial of $\sqrt{2} + \sqrt{3}$ is of degree 4. From the information that

$$(\sqrt{2} + \sqrt{3})^2 = 5 + 2\sqrt{6}, \quad (\sqrt{2} + \sqrt{3})^4 = 49 + 20\sqrt{6}$$

we deduce that the minimum polynomial is $X^4 - 10X^2 + 1$.

If α has a minimum polynomial over K, we say that α is **algebraic over** K and that $K[\alpha]$ $(= K(\alpha))$ is a **simple algebraic extension of** K. A complex number that is algebraic over $\mathbb{Q}$ is called an **algebraic number**. If $K(\alpha)$ is isomorphic to the field $K(X)$ of rational functions, we say that α is **transcendental over** K and that $K(\alpha)$ is a **simple transcendental extension of** K. A **transcendental number** α is a complex number that is transcendental over $\mathbb{Q}$.

Examples 3.9 and 3.10 feature simple algebraic extensions and elements $i\sqrt{3}$, $\sqrt{2}$, $\sqrt{3}$, $\sqrt{2} + \sqrt{3}$, all of which are algebraic numbers. So far we have not demonstrated that a simple transcendental extension exists. Well, yes, it does: if we take $L = K(X)$, the field of rational forms over X, then it is immediate from the definitions that the element X is transcendental over K. That, you might legitimately feel, is something of a technical knock-out, and leads to the more interesting question: do there exist transcendental complex numbers? The answer is yes, and the proof, which involves some knowledge of infinite cardinal numbers, is interesting. First, we make a fairly easy observation:

Theorem 3.11

Let $K(\alpha)$ be a simple transcendental extension of a field K. Then the degree of $K(\alpha)$ over K is infinite.

Proof

The elements $1, \alpha, \alpha^2, \ldots$ are linearly independent over K. $\qquad\qquad\square$

An extension L of K is said to be an **algebraic extension** if every element of L is algebraic over K. Otherwise L is a **transcendental extension**.

Theorem 3.12

Every finite extension is algebraic.

Proof

Let L be a finite extension of K, and suppose, for a contradiction, that L contains an element α that is transcendental over K. Then the elements $1, \alpha, \alpha^2, \ldots$ are linearly independent over K, and so $[L : K]$ cannot be finite. $\square$

Theorem 3.13

Let $L : K$ and $M : L$ be field extensions, and let $\alpha \in M$. If α is algebraic over K, then it is also algebraic over L.

Proof

Since α is algebraic over K, there exists a non-zero polynomial f in $K[X]$ such that $f(\alpha) = 0$. Since f is also in $L[X]$, we deduce that α is algebraic over L. $\square$

Remark 3.14

The minimum polynomial of α over L may of course be of smaller degree than the minimum polynomial over K. In Example 3.10 we saw that the minimum polynomial of $\sqrt{2} + \sqrt{3}$ over $\mathbb{Q}$ is $X^4 - 10X^2 + 1$. The minimum polynomial over $\mathbb{Q}[\sqrt{2}]$ is $X^2 - 2\sqrt{2}X - 1$. See Exercise 2.4 for its minimum polynomial over $\mathbb{Q}[\sqrt{3}]$.

Theorem 3.15

Let L be an extension of a field K, and let $\mathcal{A}(L)$ be the set of all elements in L that are algebraic over K. Then $\mathcal{A}(L)$ is a subfield of L.

Proof

Suppose that $\alpha, \beta \in \mathcal{A}(L)$. Then

$$\alpha - \beta \in K(\alpha, \beta) = \big(K[\alpha]\big)[\beta].$$

By Theorem 3.13, β is algebraic over $K[\alpha]$, and so both $[K[\alpha] : K]$ and $\big[\big(K[\alpha]\big)[\beta] : K[\alpha]\big]$ are finite. From Theorem 3.6 it follows that $[K(\alpha, \beta) : K]$ is finite, and so, by Theorem 3.12, $\alpha - \beta$ is algebraic over K. An identical argument shows that $\alpha/\beta \in \mathcal{A}(L)$ for all α and β ($\neq 0$) in $\mathcal{A}(L)$. $\square$

If we take K as the field $\mathbb{Q}$ of rational numbers and L as the field $\mathbb{C}$ of complex numbers, then $\mathcal{A}(K)$ is the field $\mathbb{A}$ of **algebraic numbers**.

Theorem 3.16

The field $\mathbb{A}$ of algebraic numbers is countable.

Proof

The proof depends on some knowledge of the arithmetic of infinite cardinal numbers. It is known (see [6]) that $\mathbb{Q}$ is countable. To put it in the standard notation for cardinal numbers, $|\mathbb{Q}| = \aleph_0$. Since $\mathbb{Q} \subseteq \mathbb{A}$, we know that $|\mathbb{A}| \geq \aleph_0$.

Now, the number of monic polynomials of degree n with coefficients in $\mathbb{Q}$ is $\aleph_0^n = \aleph_0$. Each such polynomial has at most n distinct roots in $\mathbb{C}$, and so the number of roots of monic polynomials of degree n is at most $n\aleph_0 = \aleph_0$. Hence the number of roots of monic polynomials of all possible degrees is at most $\aleph_0 . \aleph_0 = \aleph_0$. Thus $|\mathbb{A}| \leq \aleph_0$, and the result follows. $\square$

Theorem 3.17

Transcendental numbers exist.

Proof

It is known (see [6]) that $|\mathbb{R}| = |\mathbb{C}| = 2^{\aleph_0} > \aleph_0$. It follows that $\mathbb{C} \setminus \mathbb{A}$, the set of transcendental numbers, is non-empty. Indeed, since $|\mathbb{C} \setminus \mathbb{A}| = 2^{\aleph_0} > |\mathbb{A}|$, we can say that "most" complex numbers are transcendental. $\square$

Remark 3.18

This argument, due to Cantor[1], was extraordinary, in that it demonstrated the existence of transcendental numbers without producing a single example of such a number! Not everyone (see [2]) was convinced by a "non-constructive" argument of this type. (See [2].) As early as 1844, however, Liouville[2] had demonstrated that $\sum_{n=1}^{\infty} 10^{-n!}$ is transcendental. Proving that an interesting and important number is transcendental is, of course, harder. Hermite [3] proved in 1873 that e is transcendental, and in 1882 Lindemann [4] proved the transcendentality of π. (See [1].)

---

[1] Georg Ferdinand Ludwig Philipp Cantor, 1845–1918.
[2] Joseph Liouville, 1809–1882.
[3] Charles Hermite, 1822–1901.
[4] Carl Louis Ferdinand von Lindemann, 1852–1939.

Theorem 3.19

Let L be an extension of F, and let the elements $\alpha_1, \alpha_2, \ldots, \alpha_n$ of L have minimum polynomials $m_1, m_2, \ldots, m_n$, respectively, over F. Then

$$[F(\alpha_1, \alpha_2, \ldots, \alpha_n) : F] \leq \partial m_1 \, \partial m_2 \ldots \partial m_n. \tag{3.2}$$

Proof

The proof is by induction on n, it being clear that $[F(\alpha_1) : F] = \partial m_1$. Suppose inductively that

$$[F(\alpha_1, \alpha_2, \ldots \alpha_{n-1}) : F] \leq \partial m_1 \, \partial m_2 \ldots \partial m_{n-1}.$$

We know that $m_n(\alpha_n) = 0$. The element α_n is certainly algebraic over $F(\alpha_1, \alpha_2, \ldots \alpha_{n-1})$, and its minimum polynomial over that field must have degree not greater than ∂m_n. Thus

$$[F(\alpha_1, \alpha_2, \ldots, \alpha_n) : F(\alpha_1, \alpha_2, \ldots \alpha_{n-1})] \leq \partial m_n,$$

and the required result follows from Theorem 3.3.   □

Remark 3.20

We cannot assert equality in the formula (3.2). For example,

$$[\mathbb{Q}(\sqrt{2}) : \mathbb{Q}] = [\mathbb{Q}(\sqrt{3}) : \mathbb{Q}] = [\mathbb{Q}(\sqrt{6}) : \mathbb{Q}] = 2,$$

but $[\mathbb{Q}(\sqrt{2}, \sqrt{3}, \sqrt{6}) : \mathbb{Q}] = 4$.

Example 3.21

Show that an extension L of a field K is finite if and only if, for some n, there exist $\alpha_1, \alpha_2, \ldots, \alpha_n$, algebraic over K, such that $L = K(\alpha_1, \alpha_2, \ldots, \alpha_n)$.

Solution

Theorem 3.19 gives half of this result. Suppose now that $[L : K]$ is finite, and that $\{\alpha_1, \alpha_2, \ldots, \alpha_n\}$ is a basis for L over K. The elements α_i are all algebraic, by Theorem 3.12. Then L consists of linear combinations (with coefficients in K) of $\alpha_1, \alpha_2, \ldots, \alpha_n$, but in fact contains (and is thus equal to) the seemingly larger set $K(\alpha_1, \alpha_2, \ldots, \alpha_n)$.   □

EXERCISES

3.3. Show that, if n is not a perfect square, the field $\mathbb{Q}[\sqrt{n}]$ is isomorphic to the field

$$K = \left\{ \begin{pmatrix} a & b \\ nb & a \end{pmatrix} : a, b \in \mathbb{Q} \right\}.$$

Why does this fail if n is a perfect square?

3.4. For arbitrary a, b in $\mathbb{Q}$, find the minimum polynomial of $a + b\sqrt{2}$ over $\mathbb{Q}$.

3.5. Let $L : K$ be a field extension such that $[L : K] = 2$. Show that $L = K(\beta)$, where β is an arbitrarily chosen element of $L \setminus K$ and has a minimum polynomial of degree 2.

3.6. Let α be a root in $\mathbb{C}$ of the polynomial $X^2 + 2X + 5$. Express the element

$$\frac{\alpha^3 + \alpha - 2}{\alpha^2 - 3}$$

of $\mathbb{Q}(\alpha)$ as a linear combination of the basis $\{1, \alpha\}$.

3.7. Show that $f(X) = X^3 + X + 1$ is irreducible over $\mathbb{Q}$. Let α be a root of f in $\mathbb{C}$. Express

$$\frac{1}{\alpha} \quad \text{and} \quad \frac{1}{\alpha + 2}$$

as linear combinations of $\{1, \alpha, \alpha^2\}$.

3.8. In the context of Example 3.10,

(i) show that $\sqrt{3} \notin \mathbb{Q}[\sqrt{2}]$;

(ii) find the minimum polynomial of $\sqrt{2} + \sqrt{3}$ over $\mathbb{Q}[\sqrt{3}]$.

3.9. Show that $\mathbb{Q}(\sqrt{2}, \sqrt{5}) = \mathbb{Q}[\sqrt{2} + \sqrt{5}]$. Determine the minimum polynomial of $\sqrt{2} + \sqrt{5}$

(i) over $\mathbb{Q}$;   (ii) over $\mathbb{Q}[\sqrt{2}]$;   (iii) over $\mathbb{Q}[\sqrt{5}]$.

3.10. Determine the minimum polynomial over $\mathbb{Q}$ for each of

$$1 + \sqrt{3}, \ \frac{\sqrt{3}}{\sqrt{5}}, \ \sqrt{3} + \sqrt{5}, \ (1 + i)\sqrt{3}.$$

3.11. Determine the minimum polynomial of $\sqrt{1 + \sqrt{2}}$ over $\mathbb{Q}$. What is its minimum polynomial over $\mathbb{Q}[\sqrt{2}]$?

3.12. The element $1 + \sqrt{2} + \sqrt{3} + \sqrt{6}$ belongs to the field $\mathbb{Q}(\sqrt{2}, \sqrt{3})$. Compute its multiplicative inverse.

3.13. Let $L : K$ be a field extension, and let $g \in K[X]$. Show that

    (i) if g is irreducible over L, then it is irreducible over K;

    (ii) if g factorises over K then it factorises over L.

3.14. Show that there exist *real* transcendental numbers.

3.15. Let α, β be transcendental numbers. Decide whether the following conclusions are true or false:

    (i) $\mathbb{Q}(\alpha) \simeq \mathbb{Q}(\beta)$;   (ii) $\alpha\beta$ is transcendental;

    (iii) α^β is transcendental;   (iv) α^2 is transcendental.

3.16. (i) Show, by induction on n, that the determinant

$$
\Delta_n = \begin{vmatrix}
\lambda & 0 & 0 & 0 & \ldots & q_n \\
-1 & \lambda & 0 & 0 & \ldots & q_{n-1} \\
0 & -1 & \lambda & 0 & \ldots & q_{n-2} \\
0 & 0 & -1 & \lambda & \ldots & q_{n-3} \\
\vdots & \vdots & \vdots & \ddots & \ddots & \vdots \\
0 & 0 & 0 & \ldots & -1 & \lambda + q_1
\end{vmatrix}
$$

is equal to $q_n + q_{n-1}\lambda + \cdots + q_1\lambda^{n-1} + \lambda^n$.

(ii) Let α be algebraic over $\mathbb{Q}$, with minimum polynomial

$$
m(X) = X^n + a_{n-1}X^{n-1} + \cdots + a_1 X + a_0 .
$$

Let T_α be a linear mapping of $\mathbb{Q}[\alpha]$ onto itself, defined on the basis $B = \{1, \alpha, \ldots, \alpha^{n-1}\}$ by

$$
T_\alpha(\alpha^j) = \alpha^{j+1} \; (j = 0, 1, \ldots, n - 1) .
$$

Write down the matrix A of T_α relative to the basis B, and show that the determinant (the characteristic polynomial of A) $|XI_n - A|$ is equal to $m(X)$.

3.3 Polynomials and Extensions

In the last section, called Extensions and Polynomials, the main result was that every simple algebraic extension $K(\alpha)$ within a field L is associated with a polynomial, the minimum polynomial of α. We required α to exist within a field L. By changing the order of the words in the title we change the question: given a field K and a monic irreducible polynomial m with coefficients in K, can we create a field, an extension of K, containing an element α whose minimum polynomial is m?

Let K be a field, and let $m \in K[X]$ be irreducible and monic. Let $L = K[X]/\langle m \rangle$. Then L is a field, by Theorem 2.4. By Theorem 2.19, the mapping $a \mapsto a + \langle m \rangle$ is a monomorphism from K into L, and so L is an extension of K. Let $\alpha = X + \langle m \rangle$. Then, for each polynomial $f = a_0 + a_1 X + a_2 X^2 + \cdots + a_n X^n$ in $K[X]$,

$$
\begin{aligned}
f(\alpha) &= a_0 + a_1 \alpha + \cdots + a_n \alpha^n \\
&= a_0 + a_1 \big(X + \langle m \rangle \big) + a_2 \big(X + \langle m \rangle \big)^2 + \cdots + a_n \big(X + \langle m \rangle \big)^n \\
&= a_0 + a_1 \big(X + \langle m \rangle \big) + a_2 \big(X^2 + \langle m \rangle \big) + \cdots + a_n (X^n + \langle m \rangle) \\
&= (a_0 + a_1 X + a_2 X^2 + \cdots + a_n X^n) + \langle m \rangle \\
&= f + \langle m \rangle ,
\end{aligned}
$$

and so $f(\alpha) = 0 + \langle m \rangle$ if and only if $m \mid f$. Thus m is the minimum polynomial of α. We have proved the following result:

Theorem 3.22

Let K be a field and let m be a monic irreducible polynomial with coefficients in K. Then $L = K[X]/\langle m \rangle$ is a simple algebraic extension $K[\alpha]$ of K, and $\alpha = X + \langle m \rangle$ has minimum polynomial m over K.

The field L in the theorem is in effect unique:

Theorem 3.23

Let K, K' be fields, and let $\varphi : K \to K'$ be an isomorphism with canonical extension $\hat{\varphi} : K[X] \to K'[X]$. Let $f = a_n X^n + a_{n-1} X^{n-1} + \cdots + a_0$ be an irreducible polynomial of degree n with coefficients in K, and let $f' = \hat{\varphi}(f) = \varphi(a_n) X^n + \varphi(a_{n-1}) X^{n-1} + \cdots + \varphi(a_0)$. Let L be an extension of K containing a root α of f, and let L' be an extension of K' containing a root α' of f'. Then there is an isomorphism ψ from $K[\alpha]$ onto $K'[\alpha']$, an extension of φ.

Proof

The field $K[\alpha]$ consists of polynomials $b_0 + b_1 \alpha + \cdots + b_{n-1} \alpha^{n-1}$, with the obvious addition, and where multiplication is carried out using the equation

$$
\alpha^n = -\frac{1}{a_n}(a_{n-1}\alpha^{n-1} + \cdots + a_0).
$$

The mapping ψ is defined by

$$\psi(b_0 + b_1\alpha + \cdots + b_{n-1}\alpha^{n-1}) = \varphi(b_0) + \varphi(b_1)\alpha' + \cdots + \varphi(b_{n-1})(\alpha')^{n-1}.$$

In a more compact notation, we have that, for each polynomial u in $K[X]$ with $\partial u < n$,

$$\psi\big(u(\alpha)\big) = \big(\hat{\varphi}(u)\big)(\alpha').$$

It is clear that ψ is one–one and onto, and that it extends the isomorphism $\varphi : K \to K'$.

Let $u, v \in K[X]$, where $\partial u, \partial v \le n - 1$. Then it is clear that

$$\psi\big(u(\alpha) + v(\alpha)\big) = \psi\big(u(\alpha)\big) + \psi\big(v(\alpha)\big).$$

The corresponding equality for multiplication is less clear. We multiply $u(\alpha)$ and $v(\alpha)$ and use the minimum polynomial to reduce the answer to $w(\alpha)$, say, where $\partial w \le n-1$. Precisely, we use the division algorithm to write $uv = qm+w$, where $\partial w < n$. Hence

$$\psi\big(u(\alpha)v(\alpha)\big) = \psi\big(w(\alpha)\big) = \big(\hat{\varphi}(w)\big)(\alpha'). \tag{3.3}$$

The isomorphism $\hat{\varphi}$ assures us that the division algorithm in $K'[X]$ gives

$$\hat{\varphi}(u)\hat{\varphi}(v) = \hat{\varphi}(q)\hat{\varphi}(m) + \hat{\varphi}(w). \tag{3.4}$$

Hence

$$\begin{aligned}
\psi\big(u(\alpha)\big)\psi\big(v(\alpha)\big) &= \big(\hat{\varphi}(u)\big)(\alpha')\big(\hat{\varphi}(v)\big)(\alpha') \\
&= \big(\hat{\varphi}(u)\hat{\varphi}(v)\big)(\alpha') \\
&= \big(\hat{\varphi}(q)\hat{\varphi}(m) + \hat{\varphi}(w)\big)(\alpha') \quad \text{(from (3.4))} \\
&= \big(\hat{\varphi}(q)\big)(\alpha')\big(\hat{\varphi}(m)\big)(\alpha') + \big(\hat{\varphi}(w)\big)(\alpha') \\
&= \big(\hat{\varphi}(w)\big)(\alpha') \quad \text{(since } \big(\hat{\varphi}(m)\big)(\alpha') = 0).
\end{aligned}$$

Comparing this with (3.3) gives the required result. $\qquad\qquad\square$

It is worth recording as a corollary the result we obtain when K and K' are the same field:

Corollary 3.24

Let K be a field, and let f be an irreducible polynomial with coefficients in K. If L, L' are extensions of K containing roots α, α' of f, respectively, then there is an isomorphism from $K[\alpha]$ onto $K[\alpha']$ which fixes every element of K.

Since the idea will occur quite frequently as the theory develops, we shall apply the term K-**isomorphism** to an isomorphism α from L onto L' with the property that $\alpha(x) = x$ for every element of K.

Example 3.25

If $K = \mathbb{R}$ and $m = X^2 + 1$, the field $L = K[X]/\langle X^2 + 1 \rangle$ contains an element $\delta = X + \langle X^2 + 1 \rangle$ such that $\delta^2 = -1$. The polynomial $X^2 + 1$, irreducible over $\mathbb{R}$, factorises into $(X + \delta)(X - \delta)$ in the field L. Every element of L is uniquely of the form $a + b\delta$, and so L is none other than the field $\mathbb{C}$ of complex numbers.

Remark 3.26

By the fundamental theorem of algebra (see [8]) every polynomial with co-efficients in $\mathbb{C}$ factorises into linear factors. In particular, if m is irreducible in $\mathbb{Q}[X]$, then m factorises completely in $\mathbb{C}[X]$. If we know these factors, it is therefore easier and more natural to deal, for example, with the subfield $\mathbb{Q}[i\sqrt{3}] = \{a + bi\sqrt{3} : a, b \in \mathbb{Q}\}$ of $\mathbb{C}$ than with $\mathbb{Q}[X]/\langle X^2 + 3 \rangle$. The two fields are, of course, isomorphic to each other.

If, however, we are dealing, say, with extensions of $\mathbb{Z}_2$, then we are in effect obliged to carry out the more abstract procedure, as the next example shows.

Example 3.27

The polynomial $m = X^2 + X + 1$ is irreducible over $\mathbb{Z}_2$, for any proper factor would have to be either $X - 0$ or $X - 1$, and neither 0 nor 1 is a root of m. We form the field $L = \mathbb{Z}_2[X]/\langle m \rangle$. It has 4 elements, namely,

$$0 + \langle m \rangle, \ 1 + \langle m \rangle, \ X + \langle m \rangle, \ 1 + X + \langle m \rangle,$$

more conveniently written as 0, 1, α and $1 + \alpha$, where $\alpha^2 + \alpha + 1 = 0$. The addition in L is given by

$+$	0	1	α	$1+\alpha$
0	0	1	α	$1+\alpha$
1	1	0	$1+\alpha$	α
α	α	$1+\alpha$	0	1
$1+\alpha$	$1+\alpha$	α	1	0

and the multiplication by

$\cdot$	0	1	α	$1+\alpha$
0	0	0	0	0
1	0	1	α	$1+\alpha$
α	0	α	$1+\alpha$	1
$1+\alpha$	0	$1+\alpha$	1	α

Example 3.28

Show that the mapping $\varphi : \mathbb{Q}[i+\sqrt{2}] \to \mathbb{Q}[X]/\langle X^4 - 2X^2 + 9 \rangle$, defined by

$$\varphi(a) = a + \langle X^4 - 2X^2 + 9 \rangle \ (a \in \mathbb{Q}), \quad \varphi(i+\sqrt{2}) = X + \langle X^4 - 2X^2 + 9 \rangle,$$

is an isomorphism. Determine $\varphi(i)$, $\varphi(\sqrt{2})$ and $\varphi(i\sqrt{2})$.

Solution

It is clear that $\left[\mathbb{Q}[i+\sqrt{2}] : \mathbb{Q}\right] = 4$. Since

$$(i+\sqrt{2})^2 = 1 + 2i\sqrt{2} \ \text{ and } \ (i+\sqrt{2})^4 = -7 + 4i\sqrt{2},$$

the minimum polynomial of $i+\sqrt{2}$ over $\mathbb{Q}$ is $X^4 - 2X^2 + 9$. The uniqueness theorem (Theorem 3.23) implies that φ is an isomorphism.

Let $a_0, \ldots, a_3 \in \mathbb{Q}$, and observe that

$$a_0 + a_1(i+\sqrt{2}) + a_2(i+\sqrt{2})^2 + a_3(i+\sqrt{2})^3$$
$$= a_0 + a_1(i+\sqrt{2}) + a_2(1+2i\sqrt{2}) + a_3(5i-\sqrt{2})$$
$$= (a_0 + a_2) + (a_1 + 5a_3)i + (a_1 - a_3)\sqrt{2} + (2a_2)i\sqrt{2}.$$

Since $\{1, i, \sqrt{2}, i\sqrt{2}\}$ is linearly independent over $\mathbb{Q}$, this equals i if and only if

$$a_0 + a_2 = 0, \ a_1 + 5a_3 = 1, \ a_1 - a_3 = 0, \ a_2 = 0,$$

that is, if and only if $a_1 = a_3 = 1/6$ and $a_0 = a_2 = 0$. Thus

$$i = \frac{1}{6}\big((i+\sqrt{2}) + (i+\sqrt{2})^3\big)$$

and so

$$\varphi(i) = \frac{1}{6}(X + X^3) + \langle X^4 - 2X^2 + 9 \rangle.$$

In a similar way we can deduce that

$$\varphi(\sqrt{2}) = \frac{1}{6}(5X - X^3) + \langle X^4 - 2X^2 + 9 \rangle,$$
$$\varphi(i\sqrt{2}) = \frac{1}{2}(-1 + X^2) + \langle X^4 - 2X^2 + 9 \rangle.$$

$\square$

EXERCISES

3.17. Let K be a field of characteristic 0, and suppose that $X^4 - 16X^2 + 4$ is irreducible over K. Let α be the element $X + \langle X^4 - 16X^2 + 4 \rangle$ in the field $L = K[X]/\langle X^4 - 16X^2 + 4 \rangle$. Determine the minimum polynomials of α^2, $\alpha^3 - 14\alpha$, $\alpha^3 - 18\alpha$.

3.18. Show that the polynomial $X^3 + X + 1$ is irreducible over $\mathbb{Z}_2 = \{0, 1\}$, and let α be the element $X + \langle X^3 + X + 1 \rangle$ in the field $K = \mathbb{Z}_2[X]/\langle X^3 + X + 1 \rangle$. List the 8 elements of K, and show that $K \setminus \{0\}$ is a cyclic group of order 7, generated by α.

4
Applications to Geometry

4.1 Ruler and Compasses Constructions

Undoubtedly one of the early triumphs of abstract algebra was the light it shed on some classical problems of Greek mathematics, the most significant of which was referred to as "squaring the circle". This is one of very few phrases from serious mathematics to have entered the language, though a (totally unscientific) poll of non-mathematical friends suggests that its mathematical meaning is not even remotely understood. "Something to do with πr^2, is it?" is a common answer, and indeed that is correct, but it does not get to the heart of the matter.

Let us begin with some examples of **ruler and compasses constructions**. (By a ruler here we mean a straight-edge without length markings.)

Example 4.1

Let A, B be distinct points on the plane. Construct the perpendicular bisector of AB.

Solution

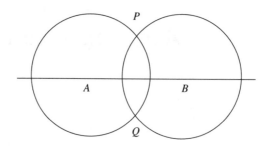

Draw the circle with centre A passing through B, and the circle with centre B passing through A. The two circles meet in points P and Q, and the line PQ is the required perpendicular bisector.   □

Example 4.2

Let A, B be distinct points on the plane, and let C be a point not on the line segment AB. Draw a line through C perpendicular to AB. (In the days when formal geometry was taught in schools, this was called **dropping a perpendicular** from C on to AB.)

Solution

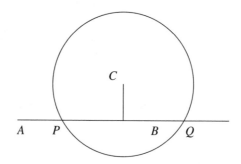

Draw a circle with centre C meeting the line AB in points P and Q. Then, as in Example 4.1, draw the perpendicular bisector of PQ.   □

Remark 4.3

This construction works just as well if C lies on the line AB.

Example 4.4

Let A, B be distinct points on the plane. Construct a square on AB.

Solution

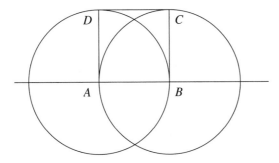

Let $\mathcal{K}_1$ be a circle with centre A passing through B, and let $\mathcal{K}_2$ be a circle with centre B passing through A. By Example 4.1 we can draw a line though A perpendicular to AB, meeting $\mathcal{K}_1$ in D, and a line through B perpendicular to AB, meeting $\mathcal{K}_2$ in C. Then $ABCD$ is the required square. □

Example 4.5

Let L be a line and A a point not on L. Construct a line through A parallel to L.

Solution

Drop a perpendicular from A on to the line L, meeting L at the point B. Then draw the perpendicular to the line AB at the point A. □

These examples are by way of preliminaries to the next, more substantial, example.

Example 4.6

Construct a square equal in area to a given rectangle.

Solution

Suppose that $AD < AB$. Draw a circle with centre A passing through D, meeting the line segment AB in E. Let M be the midpoint of AB (located using the construction in Example 4.1), and draw a circle $\mathcal{K}$ with AB as diameter. As in Example 4.2, draw the line through E perpendicular to AB, meeting the circle $\mathcal{K}$ in F. The angle AFB is a right angle, and the triangles AFB and AEF are similar. Hence

$$\frac{AE}{AF} = \frac{AF}{AB},$$

and so $AF^2 = AE.AB = AD.AB$. The square constructed on AF (by Example 4.4) has the same area as the rectangle $ABCD$.   □

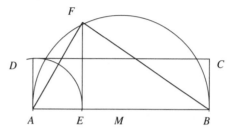

The classic challenge that intrigued mathematicians for two millennia was this:

- **squaring the circle**: to construct, using ruler and compasses only, a square equal in area to a given circle.

The problem is easily understood, and over many centuries attracted both professional mathematicians and enthusiastic amateurs. No construction was found. For a history of the problem, see [10].

Other classical challenges were

- the **duplication of the cube**: to construct a cube double the volume of a given cube;

- the **trisection of the angle**: given an angle θ, to construct the angle $\theta/3$.

EXERCISES

4.1. Show how to construct a square equal in area to a given parallelogram.

4.2. Describe a ruler and compasses construction for the bisection of an angle.

4.2 An Algebraic Approach

A cartesian coordinate system in the plane depends on

(i) specifying two axes at right angles to each other, meeting at a point O, the **origin**;

(ii) choosing a point I, distinct from O, on one of the axes, and giving it coordinates $(1, 0)$.

Let B_0 be a set of points in the plane. There are two permitted operations on the points of B_0:

(1) (Ruler) through any two points of B_0, draw a straight line;

(2) (Compasses) draw a circle whose centre is a point in B_0, and whose radius is the distance between two points in B_0.

Any point which is an intersection of two lines, or two circles, or a line and a circle, obtained by means of the operations (1) and (2), is said to be **constructed from B_0 in one step**. Denote the set of such points by $\mathcal{C}(B_0)$, and let $B_1 = B_0 \cup \mathcal{C}(B_0)$. We can continue the process, defining

$$B_n = B_{n-1} \cup \mathcal{C}(B_{n-1}) \quad (n = 1, 2, 3, \ldots). \tag{4.1}$$

A point is said to be **constructible from B_0** if it belongs to B_n for some n. A point that is constructible from $\{O, I\}$ is said to be **constructible**.

We examine Example 4.1 from this standpoint.

Example 4.7

To construct the midpoint of OI from the set $B_0 = \{O, I\}$, we carry out the following steps.

(1) Join O and I.

(2) Draw a circle with centre O, passing through I.

(3) Draw a circle with centre I, passing through O.

(4) Mark the points P, Q in which the circles intersect. Thus

$$B_1 = \{O, I, P, Q\}.$$

(5) Join P and Q.

(6) Mark the point M in which OI and PQ meet. Thus

$$B_2 = \{O, I, P, Q, M\},$$

and so the point M is constructible (from $\{O, I\}$).

This is still very geometrical. The connection with algebra comes if we associate each B_i with the subfield of $\mathbb{R}$ generated by the coordinates of the points in B_i. Let us look again at Example 4.7. As we saw in Theorem 1.15, the field K_0 generated by $B_0 = \{(0,0),(1,0)\}$ is $\mathbb{Q}$. The circles $x^2 + y^2 = 1$ and $x^2 + y^2 = 2x$ described in Steps (3) and (4) intersect in $(1/2, \pm\sqrt{3}/2)$, and so the field K_1 generated by $B_1 = \{(0,0),(1,0),(1/2,\pm\sqrt{3}/2)\}$ is $\mathbb{Q}[\sqrt{3}]$. Finally, M is the point $(1/2,0)$, and so the field K_2 generated by

$$B_2 = \{(0,0),(1,0),(\tfrac{1}{2},\pm\tfrac{1}{2}\sqrt{3}),(\tfrac{1}{2},0)\}$$

is still $\mathbb{Q}[\sqrt{3}]$. It is no accident that $[K_2 : \mathbb{Q}] = 2$:

Theorem 4.8

Let P be a constructible point, belonging (in the notation of (4.1)) to B_n, where $B_0 = \{(0,0),(1,0)\}$. For $n = 0,1,2,\ldots$, let K_n be the field generated over $\mathbb{Q}$ by B_n. Then $[K_n : \mathbb{Q}]$ is a power of 2.

Proof

It is clear that $[K_0 : \mathbb{Q}] = 1 = 2^0$. We suppose inductively that $[K_{n-1} : \mathbb{Q}] = 2^k$ for some $k \geq 0$. We require to show that $[K_n : K_{n-1}]$ is a power of 2.

New points in B_n are obtained by

(1) the intersection of two lines; or

(2) the intersection of a line and a circle; or

(3) the intersection of two circles.

Case (1) is the easiest. Suppose that we have lines AB and CD, where $A = (a_1, a_2)$, $B = (b_1, b_2)$, $C = (c_1, c_2)$, $D = (d_1, d_2)$, and that all these coordinates are in K_{n-1}. The equations of the lines are

$$(y - b_2)(a_1 - b_1) = (x - b_1)(a_2 - b_2), \quad (y - d_2)(c_1 - d_1) = (x - d_1)(c_2 - d_2),$$

and the coordinates of their intersection are obtained by solving these two simultaneous linear equations. The details are unimportant: the crucial observation is that the solution process involves only rational operations (addition, subtraction, multiplication and division), and so takes place entirely within the field K_{n-1}. The coordinates of the intersection of AB and CD lie inside the field K_{n-1}.

For Case (2), suppose that we have a line AB intersecting a circle with centre C and radius PQ, where P, Q are points with coordinates in K_{n-1}.

Taking the coordinates of A, B and C as in the previous paragraph, with all coordinates in K_{n-1}, we must solve the equations

$$(y - b_2)(a_1 - b_1) = (x - b_1)(a_2 - b_2),$$
$$(x - c_1)^2 + (y - c_2)^2 = r^2,$$

where $r^2 \in K_{n-1}$. We have to solve two simultaneous equations, one linear and one quadratic, with coefficients in K_{n-1}. Again the details are unimportant, but the standard method of doing this is to express y in terms of x using the linear equation, and then to substitute in the equation of the circle, obtaining a quadratic equation in x, with coefficients in K_{n-1}. The standard solution involves $\sqrt{\Delta}$, where Δ is the discriminant of the quadratic equation, and so the coordinates of the points of intersection belong to the field $K_{n-1}[\sqrt{\Delta}]$. (This will coincide with K_{n-1} if, by chance, $\sqrt{\Delta} \in K_{n-1}$.)

For Case (3), suppose that we have a circle with centre A and radius r and a circle with centre B with radius s, where $r, s \in K_{n-1}$. With the same notation as before, we must solve the simultaneous equations

$$(x - a_1)^2 + (y - a_2)^2 = r^2,$$
$$(x - c_1)^2 + (y - c_2)^2 = s^2.$$

By subtraction we obtain a linear equation (in fact the equation of the chord connecting the points of intersection of the circles) and so we have reduced this case to Case (2).

The conclusion is that the elements in K_n are either in K_{n-1} or in $K_{n-1}[\sqrt{\Delta}]$ for some Δ in K_{n-1}. Hence, for some $k \geq 0$,

$$K_n = K_{n-1}(\sqrt{\Delta_1}, \sqrt{\Delta_2}, \ldots, \sqrt{\Delta_k}),$$

and so $[K_n : K_{n-1}]$ is a power of 2.   □

In the light of this theorem, we now consider the three classical problems mentioned at the beginning of the chapter.

Duplicating the Cube

If (without loss of generality) we suppose that the original cube has side of length 1, we must extend the field $\mathbb{Q}$, using the construction rules, to a field K containing an element α such that $\alpha^3 = 2$. The polynomial $X^3 - 2$ is irreducible, by the Eisenstein criterion (Theorem 2.27), and so $[\mathbb{Q}[\alpha] : \mathbb{Q}] = 3$. Hence $[K : \mathbb{Q}]$ is divisible by 3, and this is impossible, by Theorem 4.8.

Trisecting the Angle

It is straightforward (see Exercise 4.2) to give a ruler and compasses construction for the bisection of a given angle. Trisection is a different story. Suppose that we have an angle 3θ, which is "known", in the sense that we know its cosine. Suppose that $\cos 3\theta = c$. We need to construct the number $\cos\theta$. Now

$$\cos 3\theta = 4\cos^3\theta - 3\cos\theta\,,$$

and so we need to find a root α of the equation $4X^3 - 3X - c = 0$. If, for example, $3\theta = \pi/2$, so that $c = 0$, then the polynomial factorises as $X(4X^2 - 3)$, and so $[\mathbb{Q}[\alpha] : \mathbb{Q}] = [\mathbb{Q}[\sqrt{3}] : \mathbb{Q}] = 2$. In this case (see Exercise 4.4) we can construct a trisector. On the other hand, if $3\theta = \pi/3$, so that $c = 1/2$, then we are looking at the polynomial $f(X) = 8X^3 - 6X - 1$. It factorises if and only if $g(X) = f(X/2) = X^3 - 3X - 1$ factorises. If $g(X)$ factorises over $\mathbb{Q}$ then it factorises over $\mathbb{Z}$, by Gauss's lemma (Theorem 2.24). One of the factors must be linear, and must be either $X - 1$ or $X + 1$. (See Example 2.26.) However, $g(1) = -3 \neq 0$ and $g(-1) = 1 \neq 0$, and so $g(X)$, and hence $f(X)$, is irreducible. Thus $[\mathbb{Q}[\alpha] : \mathbb{Q}] = 3$, and so no ruler and compasses construction is possible.

Squaring the Circle

Suppose that we have a circle of radius 1. Its area is π, and so the algebraic challenge is to construct the number $\sqrt{\pi}$. Now, as mentioned earlier, the number π is transcendental, and since $\mathbb{Q}(\pi) \subseteq \mathbb{Q}(\sqrt{\pi})$, the degree $[\mathbb{Q}(\sqrt{\pi}) : \mathbb{Q}]$ is certainly infinite. It is certainly not a power of 2, and so the construction is not possible.

This last very brief proof is of course in danger of concealing the real issue, which is the transcendentality of π. The proof of this (see [1]) is not algebraic, and would take us beyond the scope of this book. Suffice it to say that Lindemann's proof of 1882 was one of the major achievements of 19th-century mathematics.

We shall return to ruler and compasses constructions in Chapter 9.

EXERCISES

4.3. Examine the field extensions involved in the construction of the bisector of an angle.

4.4. Describe ruler and compasses constructions for the angles $\pi/3$, $\pi/4$, $\pi/6$.

Splitting Fields

When we consider a polynomial such as X^2+2 and extend the field $\mathbb{Q}$ to $\mathbb{Q}[i\sqrt{2}]$ by adjoining one of the complex roots of the polynomial, we obtain a "bonus", in that the other root $-i\sqrt{2}$ is also in the extended field. Over $\mathbb{Q}[i\sqrt{2}]$ we have that

$$X^2 + 2 = (X - i\sqrt{2})(X + i\sqrt{2}),$$

We say that the polynomial **splits completely** (into linear factors) over $\mathbb{Q}[i\sqrt{2}]$. It is indeed clear that this must happen for a polynomial of degree 2, since the "other" factor must also be linear.

By contrast, if we look at the cubic polynomial $X^3 - 2$, which is irreducible over $\mathbb{Q}$ (by the Eisenstein criterion) and if we extend $\mathbb{Q}$ to $\mathbb{Q}[\alpha]$, where $\alpha = \sqrt[3]{2}$, we obtain the factorisation

$$X^3 - 2 = (X - \alpha)(X^2 + \alpha X + \alpha^2),$$

but the quadratic factor is certainly irreducible over $\mathbb{Q}[\alpha]$. (It is indeed irreducible over $\mathbb{R}$, since the discriminant is $-3\alpha^2$.) Over the complex field we have the factorisation

$$X^3 - 2 = (X - \alpha)(X - \alpha e^{2\pi i/3})(X - \alpha e^{-2\pi i/3})$$

and, since $e^{\pm 2\pi i/3} = \frac{1}{2}(-1 \pm i\sqrt{3})$, we can say that $X^3 - 2$ splits completely over $\mathbb{Q}(\sqrt[3]{2}, i\sqrt{3})$. The degree of the extension is 6.

In general, let us consider a field K and a polynomial f in $K[X]$. We say that an extension L of K is a **splitting field** for f over K, or that $L : K$ is a **splitting field extension**, if

(i) f splits completely over L;

(ii) f does not split completely over any proper subfield E of L.

Thus, for example, $\mathbb{Q}[i\sqrt{2}]$ is a splitting field for X^2+2 over $\mathbb{Q}$, and $\mathbb{Q}(\sqrt[3]{2}, i\sqrt{3})$ is a splitting field of $X^3 - 2$ over $\mathbb{Q}$.

Theorem 5.1

Let K be a field and let $f \in K[X]$ have degree n. Then there exists a splitting field L for f over K, and $[L : K] \leq n!$.

Proof

The polynomial f has at least one irreducible factor g (which may be f itself). If, as in Theorem 3.22, we form the field $E_1 = K[X]/\langle g \rangle$ and denote the element $X + \langle g \rangle$ by α, then α has minimum polynomial g, and so $g(\alpha) = 0$. Hence g has a linear factor $Y - \alpha$ in the polynomial ring $E_1[Y]$. Moreover $[E_1 : K] = \partial g \leq n$.

We proceed inductively. Suppose that, for each r in $\{1, \ldots, n-1\}$, we have constructed an extension E_r of K such that f has at least r linear factors in $E_r[X]$, and
$$[E_r : K] \leq n(n-1)\ldots(n-r+1)\,.$$
Thus, in $E_r[X]$,
$$f = (X - \alpha_1)(X - \alpha_2)\ldots(X - \alpha_r)f_r\,,$$
and $\partial f_r = n - r$. We repeat the argument in the previous paragraph, constructing an extension E_{r+1} of E_r in which f_r has a linear factor $X - \alpha_{r+1}$ and $[E_{r+1} : E_r] \leq n - r$. We conclude that
$$[E_{r+1} : K] = [E_{r+1} : E_r]\,[E_r : K] \leq n(n-1)\ldots(n-r)\,.$$

Hence, by induction, there exists a field E_n such that f splits completely over E_n, and $[E_n : K] \leq n!$.

Now let $L = \mathbb{Q}(\alpha_1, \alpha_2, \ldots, \alpha_n) \subseteq E_n$, where $\alpha_1, \alpha_2, \ldots, \alpha_n$ (not necessarily all distinct) are the roots of f in E_n. Then f splits completely over L, and cannot split completely over any proper subfield of L. $\qquad\square$

Example 5.2

Consider the polynomial
$$f = X^5 + X^4 - X^3 - 3X^2 - 3X + 3$$

in $\mathbb{Q}[X]$, which has two irreducible factors:

$$f = (X^3 - 3)(X^2 + X - 1).$$

Let $\alpha = \sqrt[3]{3}$, and let $\gamma = (-1 + \sqrt{5})/2$, $\delta = (-1 - \sqrt{5})/2$ be the roots of $X^2 + X - 1$.

If we follow a more concrete version of the procedure in the proof of Theorem 5.1 we successively obtain

$E_1 = \mathbb{Q}(\alpha)$, $f = (X - \alpha)(X^2 + \alpha X + \alpha^2)(X^2 + X - 1)$,
$E_2 = E_1(\alpha e^{2\pi i/3})$, $f = (X - \alpha)(X - \alpha e^{2\pi i/3})(X - \alpha e^{-2\pi i/3})(X^2 + X - 1)$,
$E_3 = E_2(\alpha e^{-2\pi i/3})$, $f = (X - \alpha)(X - \alpha e^{2\pi i/3})(X - \alpha e^{-2\pi i/3})(X^2 + X - 1)$,
$E_4 = E_3(\gamma)$, $f = (X - \alpha)(X - \alpha e^{2\pi i/3})(X - \alpha e^{-2\pi i/3})(X - \gamma)(X - \delta)$,
$E_5 = E_4(\delta)$, $f = (X - \alpha)(X - \alpha e^{2\pi i/3})(X - \alpha e^{-2\pi i/3})(X - \gamma)(X - \delta)$,

where

$$[E_1 : \mathbb{Q}] = 3, \ [E_2 : E_1] = 2, \ [E_3 : E_2] = 1, \ [E_4 : E_3] = 2, \ [E_5 : E_4] = 1,$$

and so $[E_5 : \mathbb{Q}] = 12$. The field

$$E_5 = \mathbb{Q}(\alpha, \alpha e^{2\pi i/3}, \alpha e^{-2\pi i/3}, \gamma, \delta)$$

is a splitting field for f.

This is of course an unnecessarily cumbersome process when we are dealing with extensions of $\mathbb{Q}$. Once we know the roots of f in $\mathbb{C}$, it is easy to see that a splitting field for f over $\mathbb{Q}$ is $\mathbb{Q}(\sqrt[3]{3}, i\sqrt{3}, \sqrt{5})$.

We can in fact refer to *the* splitting field of a polynomial, since it is unique up to isomorphism:

Theorem 5.3

Let K and K' be fields, and let $\varphi : K \to K'$ be an isomorphism, extending to an isomorphism $\hat{\varphi} : K[X] \to K'[X]$. Let $f \in K[X]$, and let L, L' be (respectively) splitting fields of f over K and $\hat{\varphi}(f)$ over K'. Then there is an isomorphism $\varphi^* : L \to L'$ extending φ.

Proof

Suppose that $\partial f = n$ and that in $L[X]$ we have the factorisation

$$f = \alpha(X - \alpha_1)(X - \alpha_2)\ldots(X - \alpha_n),$$

where α, the leading coefficient of f, lies in K, and $\alpha_1, \alpha_2, \ldots, \alpha_n \in L$. We may suppose that, for some $m \in \{0, 1, \ldots, n\}$, the roots $\alpha_1, \alpha_2, \ldots, \alpha_m$ are not in K, and that $\alpha_{m+1}, \ldots, \alpha_n \in K$. We shall prove the theorem by induction on m.

If $m = 0$, then all the roots are in K, and so K itself is a splitting field for f. Hence, in $K'[X]$, we have

$$\hat{\varphi}(f) = \varphi(\alpha)\big(X - \varphi(\alpha_1)\big)\big(X - \varphi(\alpha_2)\big) \ldots \big(X - \varphi(\alpha_n)\big) \,;$$

thus K' is a splitting field for $\hat{\varphi}(f)$, and $\varphi^* = \varphi$.

Suppose now that $m > 0$. We make the inductive hypothesis that, for every field E and every polynomial g in $E[X]$ having fewer than m roots outside E in a splitting field L of g, every isomorphism of E can be extended to an isomorphism of L.

Our assumption that $m > 0$ implies that the irreducible factors of f in $K[X]$ are not all linear. Let f_1 be a non-linear irreducible factor of f. Then $\hat{\varphi}(f_1)$ is an irreducible factor of $\varphi(f)$ in K'. The roots of f_1 in the splitting field L are included among the roots $\alpha_1, \alpha_2, \ldots, \alpha_n$, and we may suppose, without loss of generality, that α_1 is a root of f_1. Similarly, the list $\varphi(\alpha_1), \varphi(\alpha_2), \ldots, \varphi(\alpha_n)$ of roots of $\hat{\varphi}(f)$ includes a root $\beta_1 = \varphi(\alpha_i)$ of $\hat{\varphi}(f_1)$. (We cannot assume that $i = 1$.) By Theorem 3.23, there is an isomorphism $\varphi' : K(\alpha_1) \to K'(\beta_1)$ extending φ. Since f now has fewer than m roots outside $K(\alpha_1)$, we can use the inductive hypothesis to assert the existence of an isomorphism $\varphi^* : L \to L'$ extending $\varphi' : K(\alpha_1) \to K'(\beta_1)$, and hence extending $\varphi : K \to K'$. $\square$

Example 5.4

Determine the splitting field over $\mathbb{Q}$ of the polynomial $X^4 - 2$, and find its degree over $\mathbb{Q}$.

Solution

The polynomial $X^4 - 2$ is irreducible over $\mathbb{Q}$ by the Eisenstein criterion (Theorem 2.27). Over the complex field we have the factorisation

$$X^4 - 2 = (X - \alpha)(X + \alpha)(X - i\alpha)(X + i\alpha) \,,$$

where $\alpha = \sqrt[4]{2}$, and so the splitting field of $X^4 - 2$ is $\mathbb{Q}(\alpha, i)$. The minimum polynomial of α over $\mathbb{Q}$ certainly divides $X^4 - 2$. We know, however, from the irreducibility of $X^4 - 2$ that there are no proper divisors of $X^4 - 2$ in $\mathbb{Q}[X]$, and so the minimum polynomial is $X^4 - 2$. Thus $[\mathbb{Q}(\alpha) : \mathbb{Q}] = 4$. Also, $i \notin \mathbb{Q}(\alpha)$, since $\mathbb{Q}(\alpha) \subseteq \mathbb{R}$, and so, since i is a root of $X^2 + 1$,

$$[\mathbb{Q}(\alpha, i) : \mathbb{Q}(\alpha)] = 2 \,.$$

Hence $[\mathbb{Q}(\alpha, i) : \mathbb{Q}] = 8$. $\qquad\qquad\qquad\qquad\qquad\qquad\qquad$ □

Dealing with extensions of $\mathbb{Q}$ is aided by the knowledge that every polynomial in $\mathbb{Q}$ splits completely over $\mathbb{C}$, and so the splitting field can always be presented as a subfield of $\mathbb{C}$. The next example shows that the situation in finite fields is somewhat different.

Example 5.5

In the polynomial ring $\mathbb{Z}_3[X]$ there are 9 quadratic monic polynomials. Taking $\mathbb{Z}_3$ as $\{0, 1, -1\}$, we can write these down as

$$
\begin{array}{lll}
X^2 & X^2 + 1, & X^2 - 1, \\
X^2 + X, & X^2 + X + 1, & X^2 + X - 1, \\
X^2 - X, & X^2 - X + 1 & X^2 - X - 1.
\end{array}
$$

We can test for irreducibility of these polynomials by determining whether they have roots in $\mathbb{Z}_3$. It is clear that X^2, $X^2 + X$ and $X^2 - X$ have 0 as a root, and that $X^2 - 1$ has the root 1. Also, $X^2 + X + 1$ has the root 1 and $X^2 - X + 1$ has the root -1. The remaining polynomials

$$
X^2 + 1, \quad X^2 + X - 1, \quad X^2 - X - 1
$$

are irreducible over $\mathbb{Z}_3$. The field $L = \mathbb{Z}_3[X]/\langle X^2 + 1\rangle$ contains an element $\alpha\ (= X + \langle X^2 + 1\rangle)$ such that $\alpha^2 + 1 = 0$, and in the ring $L[X]$ the polynomial $X^2 + 1$ splits completely into $(X - \alpha)(X + \alpha)$. In fact L is the splitting field for $X^2 + 1$ over $\mathbb{Z}_3$. Similarly, $\mathbb{Z}_3[X]/\langle X^2 + X - 1\rangle$ and $\mathbb{Z}_3[X]/\langle X^2 - X - 1\rangle$ are (respectively) splitting fields for $X^2 + X - 1$ and $X^2 - X - 1$. Does this mean that we have three distinct fields of order 9?

To answer this, observe that, in L (where addition takes place modulo 3 and where $\alpha^2 = -1$),

$$
(\alpha + 1)^2 + (\alpha + 1) - 1 = (\alpha^2 - \alpha + 1) + (\alpha + 1) - 1 = (-1 - \alpha + 1) + (\alpha + 1) - 1 = 0
$$

and

$$
(-\alpha + 1)^2 + (-\alpha + 1) - 1 = (-1 + \alpha + 1) + (-\alpha + 1) - 1 = 0.
$$

Hence, in $L[X]$, the polynomial $X^2 + X - 1$ factorises into

$$
\big(X - (\alpha + 1)\big)\big(X - (-\alpha + 1)\big).
$$

Thus L is also a splitting field for $X^2 + X - 1$ over $\mathbb{Z}_3$. Similarly, in $L[X]$,

$$
X^2 - X - 1 = \big(X - (\alpha - 1)\big)\big(X - (-\alpha - 1)\big),
$$

and so L is also a splitting field for $X^2 - X - 1$ over $\mathbb{Z}_3$. From Theorem 5.3 we then deduce that

$$\mathbb{Z}_3[X]/\langle X^2 + 1 \rangle \simeq \mathbb{Z}_3[X]/\langle X^2 + X - 1 \rangle \simeq \mathbb{Z}_3[X]/\langle X^2 - X - 1 \rangle \,.$$

We can even be explicit about the isomorphisms between the fields. The field $\mathbb{Z}_3[X]/\langle X^2 + X - 1 \rangle$ is generated over $\mathbb{Z}_3$ by an element $\beta \ (= X + \langle X^2 + X - 1 \rangle)$ such that $\beta^2 + \beta - 1 = 0$. The mapping that fixes the elements of $\mathbb{Z}_3$ and sends β to $\alpha + 1$ is an isomorphism from $\mathbb{Z}_3[X]/\langle X^2 + X - 1 \rangle$ onto $\mathbb{Z}_3[X]/\langle X^2 + 1 \rangle$. Similarly, $\mathbb{Z}_3[X]/\langle X^2 - X - 1 \rangle$ is generated over $\mathbb{Z}_3[X]$ by an element γ such that $\gamma^2 - \gamma - 1 = 0$, and the mapping that fixes $\mathbb{Z}_3$ and sends γ to $\alpha - 1$ is an isomorphism.

It is clear from this example that interesting things can now be said about finite fields. This is the topic of the next chapter.

EXERCISES

5.1. Determine the splitting fields over $\mathbb{Q}$ of the following polynomials, and find their degrees over $\mathbb{Q}$:

$$X^4 - 5X^2 + 6 \,, \quad X^4 - 1 \,, \quad X^4 + 1 \,.$$

5.2. Determine the splitting fields over $\mathbb{Q}$ of the following polynomials, and find their degrees over $\mathbb{Q}$:

$$X^6 - 1 \,, \quad X^6 + 1 \,, \quad X^6 - 27 \,.$$

5.3. Show that the splitting field of $X^4 + 3$ over $\mathbb{Q}$ is $\mathbb{Q}(i, \alpha\sqrt{2})$, where $\alpha = \sqrt[4]{3}$. What is its degree over $\mathbb{Q}$?

5.4. Show that the polynomial $f = X^3 + X^2 + 1$ is irreducible over $\mathbb{Z}_2$. Write down the multiplication table for the splitting field K of f over $\mathbb{Z}_2$, and determine the three linear factors of f in $K[X]$.

Finite Fields

We certainly know that finite fields exist. To summarise what we know already, from Theorem 1.14 and (1.20) we know that a finite field K has characteristic p, a prime number, and that its minimal subfield, known as its **prime** subfield, is

$$\{0_K, 1_K, 2\,(1_K), \ldots, (p-1)\,(1_K)\}\,.$$

The prime subfield is isomorphic to $\mathbb{Z}_p$, the field of integers modulo p.

Also, in Chapter 1 (Theorem 1.17 and Exercise 1.24), we established that, for all x, y in a field K of characteristic p, and for all $n \geq 1$,

$$(x \pm y)^{p^n} = x^{p^n} \pm y^{p^n}\,. \tag{6.1}$$

Using the theory developed in the intervening chapters, we can give a complete classification of finite fields. We need one preliminary idea, which applies to all fields. Let

$$f = a_0 + a_1 X + \cdots + a_n X^n$$

be a polynomial with coefficients in a field K. The **formal derivative** Df of f is defined by

$$Df = a_1 + 2a_2 X + \cdots + na_n X^{n-1}\,. \tag{6.2}$$

Although this is a formal procedure and has nothing to do with the analytic process of differentiation, the familiar formulae

$$D(kf) = k(Df), \quad D(f+g) = Df + Dg \quad (f,g \in K[X],\ k \in K) \tag{6.3}$$

and

$$D(fg) = (Df)g + f(Dg) \quad (f,g \in K[X]) \tag{6.4}$$

are still valid. (See Exercise 6.1.)

Theorem 6.1

Let f be a polynomial with coefficients in a field K, and let L be a splitting field for f over K. Then the roots of f in L are all distinct if and only if f and Df have no non-constant common factor.

Proof

Suppose first that f has a repeated root α in L, so that $f = (X - \alpha)^r g$, where $r \geq 2$. Then (see Exercises 6.1 and 6.2)

$$Df = (X - \alpha)^r (Dg) + r(X - \alpha)^{r-1} g \,,$$

and so f and Df have the common factor $X - \alpha$.

Conversely, suppose that f has no repeated roots. Then, for each root α of f in L, we have $f = (X - \alpha)g$, where $g(\alpha) \neq 0$. Hence, from (6.4),

$$Df = g + (X - \alpha)(Dg) \,,$$

and so $(Df)(\alpha) = g(\alpha) \neq 0$. Thus, by the remainder theorem (Theorem 2.18), $(X - \alpha) \nmid Df$. This holds for every factor of f in $L[X]$, and so f and Df must be coprime. $\qquad\square$

We now state the result that classifies all finite fields:

Theorem 6.2

(i) Let K be a finite field. Then $|K| = p^n$ for some prime p and some integer $n \geq 1$. Every element of K is a root of the polynomial $X^{p^n} - X$, and K is a splitting field of this polynomial over the prime subfield $\mathbb{Z}_p$.

(ii) Let p be a prime, and let $n \geq 1$ be an integer. There exists, up to isomorphism, exactly one field of order p^n.

Proof

(i) Let K have characteristic p. Then K is a finite extension of $\mathbb{Z}_p$, of degree n, say. If $\{\delta_1, \delta_2, \ldots, \delta_n\}$ is a basis of K over $\mathbb{Z}_p$, then every element of K is uniquely expressible as a linear combination

$$a_1 \delta_1 + a_2 \delta_2 + \cdots + a_n \delta_n \,,$$

with coefficients in $\mathbb{Z}_p$. For each coefficient a_i there are p choices, namely $0, 1, \ldots, p - 1$, and so there are p^n linear combinations in all. Thus $|K| = p^n$.

The group K^* is of order $p^n - 1$. Let $\alpha \in K^*$. Then, by Lagrange's theorem (Theorem 1.19), the order of α, which is the order of the subgroup $\langle \alpha \rangle$ generated by α, divides $p^n - 1$. Certainly $\alpha^{p^n - 1} = 1$. Thus $\alpha^{p^n} - \alpha = 0$ and, since we also have $0^{p^n} - 0 = 0$, we conclude that every element of K is a root of the polynomial $X^{p^n} - X$.

It follows that the polynomial $X^{p^n} - X$ splits completely over K, since $X - \alpha$ is a linear factor for each of the p^n elements α of K. It clearly cannot split completely over any proper subfield of K, and so K must be the splitting field of $X^{p^n} - X$ over $\mathbb{Z}_p$.

(ii) Let p and n be given, and let L be the splitting field of $f = X^{p^n} - X$ over $\mathbb{Z}_p$. Then, since the field is of characteristic p,

$$Df = p^n X^{p^n - 1} - 1 = -1 \,.$$

Thus f and Df are certainly coprime, and so, by Theorem 6.1, $X^{p^n} - X$ has p^n distinct roots in L. Let K be the set consisting of those roots. We show that K is a subfield of L. The elements 0, 1 are clearly in K. Suppose that $a, b \in K$. Then, by (6.1),

$$(a - b)^{p^n} = a^{p^n} - b^{p^n} = a - b \,,$$

and so $a - b \in K$. Also, if $b \neq 0$,

$$(ab^{-1})^{p^n} = a^{p^n} (b^{p^n})^{-1} = ab^{-1} \,,$$

and so $ab^{-1} \in K$. The field K is in fact itself the splitting field, since it contains (indeed consists of) all the roots of $X^{p^n} - X$, and clearly no proper subfield of K has this property.

We have shown that, for all primes p and all integers $n \geq 1$, there exists a field of order p^n. We have shown also that any field of order p^n is the splitting field of $X^{p^n} - X$ over $\mathbb{Z}_p$, and so, by Theorem 5.3, all such fields are isomorphic.
□

We have achieved a remarkably complete classification of finite fields: only fields of prime-power order exist, and in effect, for a given p and n there is exactly one field of order p^n. We call it the **Galois field** of order p^n, and denote it by $GF(p^n)$. To complete the description we need to prove one final result:

Theorem 6.3

The group of non-zero elements of the Galois field $GF(p^n)$ is cyclic.

To prove this we need some group theory. Let G be a finite group. Recall that the **order** $o(a)$ of an element a in G is the least positive integer k such that $a^k = 1$ (we are writing the identity element of G as 1) and that $a^m = 1$ if and only if $o(a)$ divides m. The **exponent** $e = e(G)$ of G is the smallest positive integer $e = e(G)$ with the property that $a^e = 1$ for all a in G. The exponent always exists (in a finite group): it is the least common multiple of the orders of the elements of G. Since $o(a)$ divides $|G|$ for every a, we can deduce that $e(G)$ divides $|G|$.

In a non-abelian group G it is possible that $o(a) < e(G)$ for all a in G. For example, in the smallest non-abelian group $S_3 = \{1, a, b, x, y, z\}$, with multiplication table

	1	a	b	x	y	z
1	1	a	b	x	y	z
a	a	b	1	z	x	y
b	b	1	a	y	z	x
x	x	y	z	1	a	b
y	y	z	x	b	1	a
z	z	x	y	a	b	1

we have $o(1) = 1$, $o(x) = o(y) = o(z) = 2$, $o(a) = o(b) = 3$, and $e(S_3) = 6$. This cannot happen, however, if the group is abelian:

Theorem 6.4

Let G be a finite abelian group with exponent e. Then there exists an element a in G such that $o(a) = e$.

Proof

Suppose that

$$e = p_1^{\alpha_1} p_2^{\alpha_2} \ldots p_k^{\alpha_k},$$

where $p_1, p_2, \ldots, p_k$ are distinct primes and $\alpha_1, \alpha_2, \ldots, \alpha_k \geq 1$. Since e is the least common multiple of the orders of the elements of G, there must exist an element h_1 whose order is divisible by $p_1^{\alpha_1}$: thus $o(h_1) = p_1^{\alpha_1} q_1$, where q_1 divides $p_2^{\alpha_2} \ldots p_k^{\alpha_k}$. Let $g_1 = h_1^{q_1}$. Then, for all $m \geq 1$, we have $g_1^m = h_1^{mq_1}$, and this is equal to 1 if and only if $p_1^{\alpha_1} q_1 \mid mq_1$, that is, if and only if $p_1^{\alpha_1} \mid m$. Thus $o(g_1) = p_1^{\alpha_1}$.

Similarly, for $i = 2, \ldots, k$, we can find an element g_i of order $p_i^{\alpha_i}$. Let

$$a = g_1 g_2 \ldots g_k,$$

and let $n = o(a)$. Thus

$$a^n = g_1^n g_2^n \ldots g_k^n = 1$$

(this is where we are using the abelian property) and so

$$g_1^n = g_2^{-n} \cdots g_k^{-n}.$$

Let $r = p_2^{\alpha_2} \ldots p_k^{\alpha_k}$. Then, since $g_i^{-nr} = 1$ for $i = 2, \ldots, k$, it follows that $g_1^{nr} = 1$. Thus $p_1^{\alpha_1}$ divides nr, and so, since p_1 and r are coprime, $p_1^{\alpha_1}$ divides n.

Similarly, $p_i^{\alpha_i}$ divides n for $i = 2, \ldots, k$, and we deduce that $e \mid n$. Since, from the definition of the exponent, we also have $n \mid e$, we deduce that $o(a) = e$. $\square$

The following corollary is immediate:

Corollary 6.5

If G is a finite abelian group such that $e(G) = |G|$, then G is cyclic.

Proof of Theorem 6.3

Denote $\mathrm{GF}(p^n)$ by K and, as usual, denote the abelian group of non-zero elements of K by K^*. Let e be the exponent of K^*. Then $a^e = 1$ for all a in K^*, and so every element of K^* is a root of the polynomial $X^e - 1$. This polynomial has at most e roots, and so $|K^*| \leq e$. But we also have $e \leq |K^*|$. Hence $e = |K^*|$ and so, by Corollary 6.5, K^* is cyclic. $\square$

Remark 6.6

Since all fields of order p^n are isomorphic, we can construct $\mathrm{GF}(p^n)$ simply by finding an irreducible polynomial f of degree n in $\mathbb{Z}_p[X]$. Then $\mathrm{GF}(p^n) = \mathbb{Z}_p[X]/\langle f \rangle$. There will, however, normally be may choices for f. See Example 5.5.

Example 6.7

Recall from Example 5.5 that the non-zero elements of the field $\mathrm{GF}(9)$ are

$$1, \quad -1, \quad \alpha, \quad 1+\alpha, \quad -1+\alpha, \quad -\alpha, \quad 1-\alpha, \quad -1-\alpha,$$

where $\alpha^2 = -1$. The orders of the elements of the group are easily computed:

$$o(1) = 1, \quad o(-1) = 2, \quad o(\pm\alpha) = 4, \quad o(\pm 1 \pm \alpha) = 8.$$

Any one of the four elements $\pm 1 \pm \alpha$ is a generator of the group. For example, the powers of $1 + \alpha$ are as listed in the table below:

n	1	2	3	4	5	6	7	8
$(1+\alpha)^n$	$1+\alpha$	$-\alpha$	$1-\alpha$	-1	$-1-\alpha$	α	$-1+\alpha$	1

EXERCISES

6.1. Let f, g be polynomials over a field K, with $\partial f = m$, $\partial g = n$.

(i) Show that $D(f + g) = Df + Dg$.

(ii) Show, by induction on $m + n$, that

$$D(fg) = (Df)g + f(Dg).$$

6.2. Show, by induction on n, that $D[(X - \alpha)^n] = n(X - \alpha)^{n-1}$.

6.3. Let p be a prime. Show that there are $p(p-1)/2$ irreducible quadratic polynomials in $\mathbb{Z}_p[X]$.

6.4. Show that $X^2 + 2$ is an irreducible polynomial over $\mathbb{Z}_5 = \{0, \pm 1, \pm 2\}$. If α is the element $X + \langle X^2 + 2 \rangle$ in the field $K = \mathrm{GF}(25) = \mathbb{Z}_5[X]/\langle X^2 + 2 \rangle$, show that $1 + \alpha$ is a generator of the 24-element cyclic group K^*.

6.5. Show that $X^4 + X + 1$ is irreducible over $\mathbb{Z}_2$. List the powers of the element $\alpha = X + \langle X^4 + X + 1 \rangle$ of $\mathbb{Z}_2[X]/\langle X^4 + X + 1 \rangle$.

6.6. Let K be a field of non-zero characteristic p.

(i) Show that the mapping $\varphi : K \to K$ given by

$$\varphi(a) = a^p \quad (a \in K)$$

is a monomorphism (called the **Frobenius[1] monomorphism**). Show (a) that this is an automorphism if the field is finite; (b) that φ is the identity map if $K = \mathbb{Z}_p$.

(ii) Give an example of an infinite K where φ does not map onto K.

6.7. With reference to Exercises 6.4 and 6.6,

(i) find the image of α under the Frobenius automorphism of $\mathrm{GF}(25)$;

(ii) in the field $\mathrm{GF}(16)$, find the image of α under φ, φ^2 and φ^3, where φ is the Frobenius automorphism.

---

[1] Ferdinand Georg Frobenius, 1849–1917.

7

The Galois Group

7.1 Monomorphisms between Fields

Mathematicians frequently draw a distinction between the theory of fields and Galois theory. The distinction is to some extent artificial, but the study of fields enters a new phase when we consider automorphisms. It is worth emphasising that the language we use (automorphisms, groups, normal subgroups, etc.) was not available to Galois. Even with the convenient language of abstract algebra, the chain of argument in this chapter is long and, at times, far from easy: the theory developed by Galois, who lacked our advantages, is surely one of the most remarkable achievements in all mathematics.

We begin with something quite general. Let K be a field, and let S be a non-empty set. Let $\mathcal{M}$ be the set of mappings from S into K. If $\theta, \varphi \in \mathcal{M}$, then $\theta + \varphi$, defined by

$$(\theta + \varphi)(s) = \theta(s) + \varphi(s) \quad (s \in S), \tag{7.1}$$

is a mapping from S into K, and so belongs to $\mathcal{M}$. Similarly, if $\theta \in \mathcal{M}$ and $a \in K$, then $a\theta$, defined by

$$(a\theta)(s) = a\theta(s) \quad (s \in S), \tag{7.2}$$

belongs to $\mathcal{M}$. It is easy to verify that $\mathcal{M}$ is a vector space with respect to these two operations. The zero vector in $\mathcal{M}$ is the mapping ζ given by

$$\zeta(s) = 0 \quad (s \in S). \tag{7.3}$$

We shall normally denote the mapping ζ simply by 0, since the context will usually make it clear whether we mean the zero element of K or the mapping ζ.

A set $\{\theta_1, \theta_2, \ldots, \theta_n\}$ of elements of $\mathcal{M}$ is linearly independent if, for all $a_1, a_2, \ldots, a_n$ in K,

$$a_1\theta_1(s) + a_2\theta_2(s) + \cdots + a_n\theta_n(s) = 0$$

for all s in S if and only if $a_1 = a_2 = \cdots = a_n = 0$. More compactly, we can write the condition as

$$a_1\theta_1 + a_2\theta_2 + \cdots + a_n\theta_n = 0 \text{ (strictly, } \zeta) \iff a_1 = a_2 = \cdots = a_n = 0.$$

The next result, due to Dedekind[1], is concerned with the case where S is itself a field. It will be one of the many important stages in the proof of the fundamental result in Section 7.6.

Theorem 7.1

Let K and L be fields, and let $\theta_1, \theta_2, \ldots, \theta_n$ be distinct monomorphisms from K into L. Then $\{\theta_1, \theta_2, \ldots, \theta_n\}$ is a linearly independent set in the vector space $\mathcal{M}$ of all mappings from K into L.

Proof

We prove the theorem by induction on n. It is clearly true for $n = 1$, since θ_1, being a monomorphism, maps the identity 1 of K to the identity 1 of L, and so is not the zero mapping defined by (7.3).

Assume now that we have established that every set of fewer than n distinct monomorphisms of K into L is linearly independent. Suppose, for a contradiction, that there exist $a_1, a_2, \ldots, a_n$ in L, not all zero, such that

$$a_1\theta_1 + a_2\theta_2 + \cdots + a_n\theta_n = 0. \tag{7.4}$$

In fact we may assume that *all* of the a_i are non-zero: if, for example, $a_n = 0$, then $\{\theta_1, \theta_2, \ldots, \theta_{n-1}\}$ is linearly dependent, in contradiction to the induction hypothesis. Dividing by a_n in (7.4) gives

$$b_1\theta_1 + \cdots + b_{n-1}\theta_{n-1} + \theta_n = 0, \tag{7.5}$$

where $b_i = a_i/a_n$ $(i = 1, 2, \ldots, n-1)$.

---

[1] Julius Wilhelm Richard Dedekind, 1831–1916.

The monomorphisms θ_1 and θ_n are by assumption distinct, and so there exists u in K such that $\theta_1(u) \neq \theta_n(u)$; the element u is certainly non-zero, as are both $\theta_1(u)$ and $\theta_n(u)$. For every z in K,

$$b_1\theta_1(uz) + \cdots + b_{n-1}\theta_{n-1}(uz) + \theta_n(uz) = 0\,, \tag{7.6}$$

and so, since $\theta_1, \theta_2, \ldots, \theta_n$ are monomorphisms,

$$b_1\theta_1(u)\theta_1(z) + \cdots + b_{n-1}\theta_{n-1}(u)\theta_{n-1}(z) + \theta_n(u)\theta_n(z) = 0\,. \tag{7.7}$$

Dividing this by $\theta_n(u)$ gives the result that, for all z in K,

$$b_1\frac{\theta_1(u)}{\theta_n(u)}\theta_1(z) + \cdots + b_{n-1}\frac{\theta_{n-1}(u)}{\theta_n(u)}\theta_{n-1}(z) + \theta_n(z) = 0\,. \tag{7.8}$$

Rewriting this as an equation concerning mappings gives

$$b_1\frac{\theta_1(u)}{\theta_n(u)}\theta_1 + \cdots + b_{n-1}\frac{\theta_{n-1}(u)}{\theta_n(u)}\theta_{n-1} + \theta_n = 0\,, \tag{7.9}$$

where the 0 on the right now stands for the zero mapping defined by (7.3). We subtract (7.9) from (7.5) and obtain

$$b_1\left(1 - \frac{\theta_1(u)}{\theta_n(u)}\right)\theta_1 + \cdots + b_{n-1}\left(1 - \frac{\theta_{n-1}(u)}{\theta_n(u)}\right)\theta_{n-1} = 0\,. \tag{7.10}$$

Our choice of u as an element such that $\theta_1(u) \neq \theta_n(u)$ means that the coefficient of θ_1 is non-zero. Thus (7.10) implies that the set $\{\theta_1, \theta_2, \ldots, \theta_{n-1}\}$ is linearly dependent, in contradiction to the induction hypothesis. $\square$

Remark 7.2

It is important to realise that the set of monomorphisms from K into L is not a subspace of the vector space $\mathcal{M}$: if θ_1 and θ_2 are monomorphisms, and if 1_K and 1_L are (respectively) the identities of K and L, then

$$(\theta_1 + \theta_2)(1_K) = \theta_1(1_K) + \theta_2(1_K) = 1_L + 1_L \neq 1_L\,,$$

and so $\theta_1 + \theta_2$ is not a monomorphism.

7.2 Automorphisms, Groups and Subfields

The first result, stated and proved for fields, applies to much more general types of algebra:

Theorem 7.3

Let K be a field. Then the set Aut K of automorphisms of K forms a group under composition of mappings.

Proof

Composition of mappings is always associative, since, for all x in K and all α, β and γ in Aut K,

$$[(\alpha \circ \beta) \circ \gamma](x) = (\alpha \circ \beta)[\gamma(x)] = \alpha\Big(\beta\big(\gamma(x)\big)\Big),$$

$$[\alpha \circ (\beta \circ \gamma)](x) = \alpha\big([\beta \circ \gamma](x)\big) = \alpha\Big(\beta\big(\gamma(x)\big)\Big).$$

There exists an identity automorphism ι in Aut K, defined by the property that $\iota(x) = x$ for all x in K, and clearly $\iota \circ \alpha = \alpha \circ \iota = \alpha$ for all α in Aut K. Finally, for every automorphism α in Aut K, there is an inverse mapping α^{-1} defined by the property that $\alpha^{-1}(x)$ is the unique z in K such that $\alpha(z) = x$. This map is also an automorphism. To see this, let $x, y \in K$, and let $\alpha^{-1}(x) = z$, $\alpha^{-1}(y) = t$; then $\alpha(z) = x$, $\alpha(t) = y$, and so $\alpha(z + t) = x + y$. Hence

$$\alpha^{-1}(x) + \alpha^{-1}(y) = z + t = \alpha^{-1}\big(\alpha(z + t)\big) = \alpha^{-1}(x + y),$$

and we can show similarly that

$$\big(\alpha^{-1}(x)\big)\big(\alpha^{-1}(y)\big) = \alpha^{-1}(xy).$$

Thus $\alpha^{-1} \in G$, and has the property that $\alpha \circ \alpha^{-1} = \alpha^{-1} \circ \alpha = \iota$. Hence G is a group. $\qquad\square$

We refer to Aut K as the **group of automorphisms** of K.

Let L be an extension of a field K. An automorphism α of L is called a K-**automorphism** if $\alpha(x) = x$ for every x in K. The set of all K-automorphisms of L is denoted by $\mathrm{Gal}(L : K)$ and is called the **Galois group of L over K**. The **Galois group** $\mathrm{Gal}(f)$ **of a polynomial** f in $K[X]$ is defined as $\mathrm{Gal}(L : K)$, where L is a splitting field of f over K. The Galois group is the key to the connection between classical algebra, dominated by the theory of equations, and modern abstract algebra, and this chapter is devoted to establishing the properties that make it such an important idea. First, we hasten to justify the use of the word "group":

Theorem 7.4

Let $L : K$ be a field extension. Then the set $\mathrm{Gal}(L : K)$ of all K-automorphisms of L is a subgroup of $\mathrm{Aut}\, L$.

Proof

Certainly $\iota \in \mathrm{Gal}(L : K)$. Let $\alpha, \beta \in \mathrm{Gal}(L : K)$. Then, for all x in K,

$$x = \beta^{-1}\big(\beta(x)\big) = \beta^{-1}(x)\,,$$

and so

$$\alpha\big(\beta^{-1}(x)\big) = \alpha(x) = x\,.$$

Thus $\alpha\beta^{-1} \in \mathrm{Gal}(L : K)$, and so, by (1.23), $\mathrm{Gal}(L : K)$ is a subgroup of $\mathrm{Aut}\, L$. $\qquad\square$

We now introduce an important idea connecting the subfields E of L containing K and the subgroups H of the group $\mathrm{Gal}(L : K)$. For each E we define

$$\Gamma(E) = \{\alpha \in \mathrm{Aut}\, L : \alpha(z) = z \text{ for all } z \text{ in } E\}\,; \qquad (7.11)$$

and for each H we define

$$\Phi(H) = \{x \in L : \alpha(x) = x \text{ for all } \alpha \text{ in } H.\} \qquad (7.12)$$

The essence of Galois theory is contained in these two mappings, and the principal thrust of this chapter is to find conditions under which they are mutually inverse. There are many technicalities involved in obtaining these conditions, but these must not obscure the final goal, which is Theorem 7.34. The technicalities concern the properties of the extension $L : K$ that will make the maps Γ and Φ mutually inverse. We require the extension to be "normal" and "separable", and these two notions are explored in Sections 7.3 and 7.4.

The following property is easily established:

Theorem 7.5

Let $L : K$ be a field extension.

(i) For every subfield E of L containing K, the set $\Gamma(E)$ is a subgroup of $\mathrm{Gal}(L : K)$.

(ii) For every subgroup H of $\mathrm{Gal}(L : K)$, the set $\varPhi(H)$ is a subfield of L containing K.

Proof

(i) Certainly $\varGamma(E)$ is non-empty, since it contains ι, the identity automorphism. Also, $\varGamma(E) \subseteq \mathrm{Gal}(L : K)$, since every automorphism fixing all elements of E automatically fixes all elements of K.

Let $\alpha, \beta \in \varGamma(E)$. Then, for all z in E,

$$\alpha\big(\beta^{-1}(z)\big) = \alpha\Big(\beta^{-1}\big(\beta(z)\big)\Big) = \alpha(z) = z \,,$$

and so $\alpha\beta^{-1} \in \varGamma(E)$. Hence, by (1.23), $\varGamma(E)$ is a subgroup.

(ii) It is clear that $K \subseteq \varPhi(H)$, since *every* automorphism in $\mathrm{Gal}(L : K)$ fixes the elements of K. Let $x, y \in \varPhi(H)$. Then, for all α in H,

$$\alpha(x - y) = \alpha(x) - \alpha(y) = x - y \,,$$

and so $x - y \in \varPhi(H)$. If $y \neq 0$, then, for all α in H,

$$\begin{aligned} \alpha(xy^{-1}) &= \alpha(x)\alpha(y^{-1}) = \alpha(x)\big(\alpha(y)\big)^{-1} \quad \text{(see Exercise 7.1)}\\ &= xy^{-1} \,, \end{aligned}$$

and so $xy^{-1} \in \varPhi(H)$. Thus $\varPhi(H)$ is a subfield of L. $\square$

At this point we have established a two-way connection between subfields of L containing K and subgroups of the group $\mathrm{Gal}(L : K)$. It is an "order-reversing" connection:

Theorem 7.6

Let $L : K$ be a field extension.

(i) If E_1 and E_2 are subfields of L containing K, then

$$E_1 \subseteq E_2 \;\Rightarrow\; \varGamma(E_1) \supseteq \varGamma(E_2) \,.$$

(ii) If H_1 and H_2 are subgroups of $\mathrm{Gal}(L : K)$, then

$$H_1 \subseteq H_2 \;\Rightarrow\; \varPhi(H_1) \supseteq \varPhi(H_2) \,.$$

Proof

(i) Suppose that $E_1 \subseteq E_2$, and let $\alpha \in \Gamma(E_2)$. Then α fixes every element of E_2 and so certainly fixes every element of E_1. Hence $\alpha \in \Gamma(E_1)$.

(ii) Suppose that $H_1 \subseteq H_2$, and let $z \in \Phi(H_2)$. Then $\alpha(z) = z$ for every α in H_2, and so certainly for every α in H_1. Hence $z \in \Phi(H_1)$. $\square$

The next natural question is concerned with whether the two mappings Γ and Φ are mutually inverse. In fact they need not be, as the following example shows.

Example 7.7

Consider the extension $\mathbb{Q}(u)$ of $\mathbb{Q}$, where $u = \sqrt[3]{2}$. If $\alpha \in \mathrm{Gal}\big(\mathbb{Q}(u) : \mathbb{Q}\big)$, then

$$\big(\alpha(u)\big)^3 = \alpha(u^3) = \alpha(2) = 2$$

and so, being real, $\alpha(u)$ must be equal to u. It follows that $\mathrm{Gal}\big(\mathbb{Q}(u) : \mathbb{Q}\big)$ is the trivial group $\{\iota\}$. Now, two mappings can be mutually inverse only if they are both bijections, and here we have $\Gamma(\mathbb{Q}(u)) = \Gamma(\mathbb{Q}) = \{\iota\}$. To look at it another way, we have

$$\Phi\big(\Gamma(\mathbb{Q})\big) = \Phi\big(\{\iota\}\big) = \mathbb{Q}(u).$$

Other examples have the desired property.

Example 7.8

Describe the group $\mathrm{Gal}(\mathbb{C} : \mathbb{R})$.

Solution

If $\alpha \in \mathrm{Gal}(\mathbb{C} : \mathbb{R})$, then $\alpha(x) = x$ for all x in $\mathbb{R}$. Let $\alpha(i) = j$. Then

$$j^2 = \big(\alpha(i)\big)^2 = \alpha(i^2) = \alpha(-1) = -1 \,,$$

and so $j = \pm i$. If $j = i$ then, for all $x + yi$ in $\mathbb{C}$ (with x, y in $\mathbb{R}$)

$$\alpha(x + yi) = \alpha(x) + \alpha(y)\alpha(i) = x + yi$$

and so $\alpha = \iota$, the identity automorphism. If $j = -i$ then $\alpha(x+yi) = x - yi$. This mapping certainly fixes the elements of $\mathbb{R}$. To check that it is an automorphism, note that

$$\alpha\big((x + yi) + (u + vi)\big) = \alpha\big((x + u) + (y + v)i\big) = (x + u) - (y + v)i$$
$$= (x - yi) + (u - vi) = \alpha(x + yi) + \alpha(u + vi) \,,$$

and

$$\alpha\big((x+yi)\,(u+vi)\big) = \alpha\big((xu-yv)+(xv+yu)i\big) = (xu-yv)-(xv+yu)i$$
$$= (x-yi)\,(u-vi) = \big(\alpha(x+yi)\big)\big(\alpha(u+vi)\big).$$

We deduce that $\mathrm{Gal}(\mathbb{C} : \mathbb{R})$ is the group $\{\iota, \kappa\}$ of order 2, where κ is the **complex conjugation** mapping sending $x+yi$ to $x-yi$. Since $[\mathbb{C} : \mathbb{R}] = 2$, a prime number, there cannot be any subfields of $\mathbb{C}$ lying between $\mathbb{C}$ and $\mathbb{R}$. We have

$$\Phi(\{\iota\}) = \mathbb{C}, \quad \Phi(\{\iota, \kappa\}) = \mathbb{R}.$$

$\square$

Before considering another example, we note that the argument above leading to the conclusion that $\alpha(i) = \pm i$ is a special case of a much more general observation as follows:

Theorem 7.9

Let K be a field, let L be an extension of K, and let $z \in L \setminus K$. If z is a root of a polynomial f with coefficients in K, and if $\alpha \in \mathrm{Gal}(L : K)$, then $\alpha(z)$ is also a root of f.

Proof

Let $f = a_0 + a_1 X + \cdots + a_n X^n$, where $a_0, a_1, \ldots, a_n \in K$, and suppose that $f(z) = 0$. Then

$$f\big(\alpha(z)\big) = a_0 + a_1\alpha(z) + \ldots + a_n\big(\alpha(z)\big)^n$$
$$= \alpha(a_0) + \alpha(a_1)\alpha(z) + \ldots + \alpha(a_n)\alpha(z^n)$$
$$= \alpha(a_0 + a_1 z + \cdots + a_n z^n)$$
$$= \alpha(0) = 0.$$

$\square$

Example 7.10

Describe the group $\mathrm{Gal}[\mathbb{Q}(\sqrt{2}, i\sqrt{3}) : \mathbb{Q}]$. For each of its subgroups H, determine $\Phi(H)$.

Solution

The elements of $\mathbb{Q}(\sqrt{2}, i\sqrt{3})$ are of the form $a + b\sqrt{2} + ci\sqrt{3} + di\sqrt{6}$. By Theorem 7.9, if $\alpha \in \mathrm{Gal}\big(\mathbb{Q}(\sqrt{2}, i\sqrt{3}), \mathbb{Q}\big)$, then $\alpha(\sqrt{2}) = \pm\sqrt{2}$, $\alpha(i\sqrt{3}) = \pm i\sqrt{3}$. There are four elements in $\mathrm{Gal}\big(\mathbb{Q}(\sqrt{2}, i\sqrt{3}), \mathbb{Q}\big)$, namely, ι, τ, θ and β, where ι is the identity map, and

- $\tau(a + b\sqrt{2} + ci\sqrt{3} + di\sqrt{6}) = a - b\sqrt{2} + ci\sqrt{3} - di\sqrt{6}$;
- $\theta(a + b\sqrt{2} + ci\sqrt{3} + di\sqrt{6}) = a + b\sqrt{2} - ci\sqrt{3} - di\sqrt{6}$;
- $\beta(a + b\sqrt{2} + ci\sqrt{3} + di\sqrt{6}) = a - b\sqrt{2} - ci\sqrt{3} + di\sqrt{6}$.

One may verify that these are all $\mathbb{Q}$-automorphisms of $\mathbb{Q}(\sqrt{2}, i\sqrt{3})$. See Exercise 7.4.

The group has multiplication table

	ι	τ	θ	β
ι	ι	τ	θ	β
τ	τ	ι	β	θ
θ	θ	β	ι	τ
β	β	θ	τ	ι

The proper subgroups of this group are $H_1 = \{\iota, \tau\}$, $H_2 = \{\iota, \theta\}$ and $H_3 = \{\iota, \beta\}$; and

$$\Phi(H_1) = \mathbb{Q}(i\sqrt{3}), \quad \Phi(H_2) = \mathbb{Q}(\sqrt{2}), \quad \Phi(H_3) = \mathbb{Q}(i\sqrt{6}).$$

It is not perhaps completely obvious that there are no other subfields of $\mathbb{Q}(\sqrt{2}, i\sqrt{3})$, but this will emerge as a consequence of the theory we shall develop. $\qquad\square$

The mappings Φ and Γ, together known as the **Galois correspondence**, need not be mutually inverse, but they do have this weaker property:

Theorem 7.11

Let L be an extension of a field K, let E be a subfield of L containing K, and let H be a subgroup of $\mathrm{Gal}(L : K)$. Then

$$E \subseteq \Phi\big(\Gamma(E)\big), \quad H \subseteq \Gamma\big(\Phi(H)\big).$$

Proof

Let $z \in E$. The group $\Gamma(E)$ is the set of all automorphisms fixing each element of E, and so z is fixed by all the automorphisms in $\Gamma(E)$. That is, $z \in \Phi\big(\Gamma(E)\big)$. Hence $E \subseteq \Phi\big(\Gamma(E)\big)$.

Let $\alpha \in H$. The field $\Phi(H)$ is the set of elements of L fixed by every element of H, and so α fixes every element of $\Phi(H)$. That is, $\alpha \in \Gamma\big(\Phi(H)\big)$. Hence $H \subseteq \Gamma\big(\Phi(H)\big)$. $\qquad\square$

Since we are dealing with *finite* extensions and *finite* groups, it would follow, for example, that $E = \Phi\big(\Gamma(E)\big)$, if we could show that $|E| = \big|\Phi\big(\Gamma(E)\big)\big|$. Results concerning cardinalities of sets are therefore relevant to our goal. We end this section with one such result. The proof is longer than one might have expected – or hoped.

Theorem 7.12

Let L be a finite extension of a field K, and let G be a finite subgroup of $\mathrm{Gal}(L : K)$. Then $[L : \Phi(G)] = |G|$.

Proof

To prove this we need to recall some standard linear algebra. (See [3].) Let V and W be finite-dimensional vector spaces over a field K, with dimensions m, n, respectively, and let $T : V \to W$ be a linear mapping. The **image** $\mathrm{im}\, T$ of T is the set $\{T(v) : v \in V\}$. It is a subspace of W, and its dimension $\dim(\mathrm{im}\, T)$ is called the **rank** $\rho(T)$ of T. The **kernel** $\ker T$ of T is the set $\{v \in V : T(v) = 0\}$. It is a subspace of V, and its dimension $\dim(\ker T)$ is called the **nullity** $\nu(T)$ of T. A standard result in linear algebra states that

$$\rho(T) + \nu(T) = \dim V = m. \tag{7.13}$$

If $n < m$, then certainly $\rho(T) \leq n < m$, and so $\nu(T) > 0$. Thus there exists a non-zero vector v in V such that $T(v) = 0$.

In more concrete terms, if we have an $n \times m$ matrix $A = [a_{ij}]_{n \times m}$ with entries in K, and an m-dimensional column vector $\mathbf{v}$, the map $\mathbf{v} \mapsto A\mathbf{v}$ is a linear mapping from the vector space K^m into the vector space K^n. From the final sentence of the last paragraph we deduce that, if $n < m$, then there exists a non-zero vector $\mathbf{v}$ such that $A\mathbf{v} = \mathbf{0}$. That is, there exist $v_1, v_2, \ldots, v_m$ in K, not all zero, such that

$$a_{1j}v_1 + a_{2j}v_2 + \cdots + a_{mj}v_m = 0 \quad (j = 1, 2, \ldots, n). \tag{7.14}$$

We are now ready to prove the statement of the theorem. Let $|G| = m$ and $[L : \Phi(G)] = n$. We show first that the statement $m > n$ leads to a contradiction, using the piece of linear algebra above.

So suppose that $m > n$, and write $G = \{\alpha_1 = \iota, \alpha_2, \ldots, \alpha_m\}$, where ι is the identity map, and suppose that $\{z_1, z_2, \ldots, z_n\}$ is a basis for L over $\Phi(G)$.

Consider the $n \times m$ matrix

$$\begin{bmatrix} \alpha_1(z_1) & \alpha_2(z_1) & \cdots & \alpha_m(z_1) \\ \alpha_1(z_2) & \alpha_2(z_2) & \cdots & \alpha_m(z_2) \\ \vdots & \vdots & & \vdots \\ \alpha_1(z_n) & \alpha_2(z_n) & \cdots & \alpha_m(z_n) \end{bmatrix}.$$

From (7.14) we deduce that there exist $v_1, v_2, \ldots, v_m$ in L, not all zero, such that

$$\alpha_1(z_j)v_1 + \alpha_2(z_j)v_2 + \cdots + \alpha_m(z_j)v_m = 0 \quad (j = 1, 2, \ldots, n). \qquad (7.15)$$

Let $b \in L$. We are supposing that $\{z_1, z_2, \ldots, z_n\}$ is a basis for L over $\Phi(G)$, and so there exist elements $b_1, b_2, \ldots, b_n$ of $\Phi(G)$ such that

$$b = b_1 z_1 + b_2 z_2 + \cdots + b_n z_n. \qquad (7.16)$$

Multiplying the n equations (7.15) by $b_1, b_2, \ldots, b_n$ (respectively) gives

$$b_j \alpha_1(z_j)v_1 + b_j \alpha_2(z_j)v_2 + \cdots + b_j \alpha_m(z_j)v_m = 0 \quad (j = 1, 2, \ldots, n). \qquad (7.17)$$

Now recall that, since the b_j all lie in $\Phi(G)$ and the α_i all lie in G, we have $b_j = \alpha_i(b_j)$ for all i and j. Thus we may rewrite the equations (7.17) as

$$\alpha_1(b_j z_j)v_1 + \alpha_2(b_j z_j)v_2 + \cdots + \alpha_m(b_j z_j)v_m = 0 \quad (j = 1, 2, \ldots, n). \qquad (7.18)$$

If we add these n equations together, and make use of (7.16), we obtain

$$v_1 \alpha_1(b) + v_2 \alpha_2(b) + \cdots + v_m \alpha_m(b) = 0.$$

This holds for all b in L, and so the automorphisms $\alpha_1, \alpha_2, \ldots, \alpha_m$ are linearly dependent. By Theorem 7.1, this is impossible. Hence $n \geq m$.

Next, suppose that $n = [L : \Phi(G)] > m$. Again we use linear algebra. This time we have subset $\{z_1, z_2, \ldots, z_{m+1}\}$ of L which is linearly independent over $\Phi(G)$, and we consider the $m \times (m+1)$ matrix

$$\begin{bmatrix} \alpha_1(z_1) & \alpha_1(z_2) & \cdots & \alpha_1(z_{m+1}) \\ \alpha_2(z_1) & \alpha_2(z_2) & \cdots & \alpha_2(z_{m+1}) \\ \vdots & \vdots & & \vdots \\ \alpha_m(z_1) & \alpha_m(z_2) & \cdots & \alpha_m(z_{m+1}) \end{bmatrix}.$$

By (7.14), there exist $u_1, u_2, \ldots, u_{m+1}$ in L, not all zero, such that

$$\alpha_j(z_1)u_1 + \alpha_j(z_2)u_2 + \cdots + \alpha_j(z_{m+1})u_{m+1} = 0 \quad (j = 1, 2, \ldots, m).$$

Let us suppose that the elements $u_1, u_2, \ldots, u_{m+1}$ are chosen *so that as few as possible are non-zero*. We may relabel the elements so that $u_1, u_2, \ldots, u_r$ are non-zero, and $u_{r+1} = \cdots = u_{m+1} = 0$. So now we have

$$\alpha_j(z_1)u_1 + \alpha_j(z_2)u_2 + \cdots + \alpha_j(z_r)u_r = 0 \quad (j = 1, 2, \ldots, m). \tag{7.19}$$

Dividing (7.19) by u_r gives a modified set of m equations

$$\alpha_j(z_1)u_1' + \cdots + \alpha_j(z_{r-1})u_{r-1}' + \alpha_j(z_r) = 0 \quad (j = 1, 2, \ldots, m), \tag{7.20}$$

where $u_i' = u_i/u_r$ $(i = 1, 2, \ldots, r-1)$. We defined α_1 to be the identity of G, and so the first of these equations is

$$z_1 u_1' + \cdots + z_{r-1} u_{r-1}' + z_r = 0. \tag{7.21}$$

If all of the elements $u_1', \ldots, u_{r-1}'$ belonged to $\Phi(G)$, then $\{z_1, z_2, \ldots, z_r\}$ would be linearly dependent over $\Phi(G)$, and we know that this is not so. Hence at least one of $u_1', \ldots, u_{r-1}'$ does not belong to $\Phi(G)$: without loss of generality, we may suppose that $u_1' \notin \Phi(G)$. That is, u_1' is not fixed by every automorphism in G, and so there is an automorphism in G, which we may take to be α_2, such that

$$\alpha_2(u_1') \neq u_1'. \tag{7.22}$$

We apply α_2 to the equations (7.21): for $j = 1, 2, \ldots, m$,

$$(\alpha_2\alpha_j)(z_1)\alpha_2(u_1') + \cdots + (\alpha_2\alpha_j)(z_{r-1})\alpha_2(u_{r-1}') + \alpha_2\alpha_j(z_r) = 0. \tag{7.23}$$

Now, since G is a group, the set $\{\alpha_2\alpha_1, \alpha_2\alpha_2, \ldots, \alpha_2\alpha_m\}$ is the same as the set $\{\alpha_1, \alpha_2, \ldots, \alpha_m\}$: only the order of the elements is different. Hence we may change the order of the listed equations (7.23) and obtain

$$\alpha_j(z_1)\alpha_2(u_1') + \cdots + \alpha_j(z_{r-1})\alpha_2(u_{r-1}') + \alpha_j(z_r) = 0 \quad (j = 1, 2, \ldots, m). \tag{7.24}$$

Subtracting (7.24) from (7.20) gives, for $j = 1, 2, \ldots, m$,

$$\alpha_j(z_1)\big(u_1' - \alpha_2(u_1')\big) + \cdots + \alpha_j(z_{r-1})\big(u_{r-1}' - \alpha_2(u_{r-1}')\big) = 0. \tag{7.25}$$

Let $v_i = u_i' - \alpha_2(u_i')$ for $i = 1, 2, \ldots, r-1$ and $v_i = 0$ for $i = r, r+1, \ldots, m+1$. Then (7.25) becomes

$$\alpha_j(z_1)v_1 + \cdots + \alpha_j(z_2)v_2 + \cdots + \alpha_j(z_{m+1})v_{m+1} = 0 \quad (j = 1, 2, \ldots, m).$$

From (7.22) we know that the elements v_i are not all zero, and we have arranged that no more than $r-1$ of the v_i are non-zero. This is a contradiction to the stated property of the elements $u_1, u_2, \ldots, u_{m+1}$, and so we conclude that it is not possible to have $[L : \Phi(G)] > m$. Hence $[L : \Phi(G)] = m$. $\qquad\square$

EXERCISES

7.1. Let α be an automorphism of a field K. Show that

  (i) $\alpha(0) = 0$, and $\alpha(-x) = -(\alpha(x))$ for all x in K;

  (ii) $\alpha(1) = 1$, and $(\alpha(x))^{-1} = \alpha(x^{-1})$ for all $x \neq 0$ in K.

7.2. Determine $\operatorname{Aut} \mathbb{Q}$ and $\operatorname{Aut} \mathbb{Z}_p$.

7.3. Show that $\Gamma \Phi \Gamma = \Gamma$ and $\Phi \Gamma \Phi = \Phi$.

7.4. Verify that the mapping τ defined in Example 7.10 is a $\mathbb{Q}$-automorphism of $\mathbb{Q}(\sqrt{2}, i\sqrt{3})$

7.5. Describe the Galois group $\operatorname{Gal}(\mathbb{Q}(i + \sqrt{2}) : \mathbb{Q})$.

7.6. Describe the Galois group $\operatorname{Gal}(\operatorname{GF}(8) : \mathbb{Z}_2)$.

7.3 Normal Extensions

In the next two sections, with a view to establishing the conditions under which the maps Γ and Φ studied in the last section are mutually inverse, we introduce two new ideas. Among the examples we have considered are two extensions of $\mathbb{Q}$, namely, $\mathbb{Q}(\sqrt{2})$ and $\mathbb{Q}(\sqrt[3]{2})$. In the first case $X^2 - 2$, the minimum polynomial of $\sqrt{2}$, splits completely over $\mathbb{Q}(\sqrt{2})$; in the second case we see that $X^3 - 2$, the minimum polynomial of $\sqrt[3]{2}$, does not split completely over $\mathbb{Q}(\sqrt[3]{2})$. This is an important difference. However, although it is convenient at times to consider arbitrary extensions $L : K$, our primary interest is with Galois groups of *polynomials*, when L is a splitting field over K for some polynomial. We shall certainly achieve this closer focus if we suppose that $L : K$ is a **normal** extension, by which we mean that every irreducible polynomial in $K[X]$ having at least one root in L splits completely over L. On the face of it this is a very strong property, and indeed it is not immediately clear that even $\mathbb{Q}(\sqrt{2})$ is a normal extension of $\mathbb{Q}$. However, we have the following result:

Theorem 7.13

A finite extension L of a field K is normal if and only if it is a splitting field for some polynomial in $K[X]$.

Proof

One way round this is fairly straightforward. Suppose that L is a finite normal extension, and let $\{z_1, z_2, \ldots, z_n\}$ be a basis for L over K. For $i = 1, 2, \ldots, n$, let m_i be the minimum polynomial of z_i, and let $m = m_1 m_2 \ldots m_n$. Each m_i has at least one root z_i in L and so splits completely over L. Hence m splits completely over L. Moreover, since L is generated by $z_1, z_2, \ldots, z_n$, it is not possible for m to split completely over any proper subfield of L. Thus L is a splitting field for m over K.

We turn now to the more surprising converse result. Suppose that E is a splitting field for some polynomial g over K, and let f, with degree at least 2, be an arbitrarily chosen irreducible polynomial in $K[X]$, having a root α in E. We must show that f splits completely over E. The polynomial fg certainly lies in $E[X]$, and has a splitting field L containing E. Suppose that β is another root of f in L. We have subfields of L as indicated in the following diagram, in which the arrows denote inclusion:

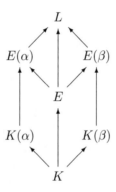

Now

$$[E(\alpha) : E]\,[E : K] = [E(\alpha) : K] = [E(\alpha) : K(\alpha)]\,[K(\alpha) : K], \qquad (7.26)$$

and

$$[E(\beta) : E]\,[E : K] = [E(\beta) : K] = [E(\beta) : K(\beta)]\,[K(\beta) : K]. \qquad (7.27)$$

Since α and β are roots of the same irreducible polynomial f, it follows from Corollary 3.24 that there is a K-isomorphism φ from $K(\alpha)$ onto $K(\beta)$. Certainly

$$[K(\alpha) : K] = [K(\beta) : K]. \qquad (7.28)$$

Since E is a splitting field for g over K, it follows that $E(\alpha)$ is a splitting field for g over $K(\alpha)$ and $E(\beta)$ is a splitting field for g over $K(\beta)$. Hence, by

Theorem 5.3, there is an isomorphism φ^* from $E(\alpha)$ onto $E(\beta)$, extending the K-isomorphism φ from $K(\alpha)$ onto $K(\beta)$. It follows in particular that

$$[E(\alpha) : K(\alpha)] = [E(\beta) : K(\beta)]. \qquad (7.29)$$

Now $[E(\alpha) : E] = 1$, since $\alpha \in E$ by assumption. Hence

$$
\begin{aligned}
[E(\beta) : E]\,[E : K] &= [E(\beta) : K(\beta)]\,[K(\beta) : K] \quad \text{(by (7.27))} \\
&= [E(\alpha) : K(\alpha)]\,[K(\alpha) : K] \quad \text{(by (7.28) and (7.29))} \\
&= [E(\alpha) : E]\,[E : K] \quad \text{(by (7.26))} \\
&= [E : K].
\end{aligned}
$$

Thus $[E(\beta) : E] = 1$ and so $\beta \in E$, as required. $\square$

Two corollaries are worth recording at this stage:

Corollary 7.14

Let L be a normal extension of finite degree over a field K, and let E be a subfield of L containing K. Then every K-monomorphism from E into L can be extended to a K-automorphism of L.

Proof

Let φ be a K-monomorphism from E into L. By Theorem 7.13, there exists a polynomial f such that L is a splitting field for f over K. It is also a splitting field for f over each of the fields E and $\varphi(E)$. By Theorem 5.3 (with $L' = L$), we deduce that there is a K-automorphism φ^* of L extending φ. $\square$

Example 7.15

Let $K = \mathbb{Q}$, $E = \mathbb{Q}(\sqrt{2})$, $L = \mathbb{Q}(\sqrt{2}, \sqrt{5})$. Let φ be the K-monomorphism from E to L defined by $\varphi(a + b\sqrt{2}) = a - b\sqrt{2}$. Then φ extends to a $\mathbb{Q}$-automorphism φ^* of L, given by

$$\varphi^*(a + b\sqrt{2} + c\sqrt{5} + d\sqrt{10}) = a - b\sqrt{2} + c\sqrt{5} - d\sqrt{10}.$$

Corollary 7.16

Let L be a normal extension of finite degree over a field K. If z_1 and z_2 are roots in L of an irreducible polynomial in $K[X]$, then there exists a K-automorphism θ of L such that $\theta(z_1) = z_2$.

Proof

By Theorem 3.24, there is a K-isomorphism from $K(z_1)$ onto $K(z_2)$. By Corollary 7.14, this extends to a K-automorphism θ of L. $\square$

Example 7.17

Let $K = \mathbb{Q}$ and let $L = \mathbb{Q}(u, i\sqrt{3})$, where $u = \sqrt[3]{2}$. Then L, being the splitting field over $\mathbb{Q}$ of $X^3 - 2$, is a normal extension. The set $\{1, u, u^2, i\sqrt{3}, ui\sqrt{3}, u^2 i\sqrt{3}\}$ is a basis for L over $\mathbb{Q}$. Consider the two roots u and $ue^{2\pi i/3}$ of the polynomial $X^3 - 2$, which is certainly irreducible over $\mathbb{Q}$. There is a $\mathbb{Q}$-isomorphism θ : $\mathbb{Q}(u) \to \mathbb{Q}(ue^{2\pi i/3})$, and by Corollary 7.16 this extends to a $\mathbb{Q}$- automorphism θ^* of L. Any $\mathbb{Q}$-automorphism of L maps $i\sqrt{3}$ to $\pm i\sqrt{3}$. If we choose $\theta^*(i\sqrt{3}) = i\sqrt{3}$, then, recalling that $e^{2\pi i/3} = \frac{1}{2}(-1 + i\sqrt{3})$, we deduce that

$$\theta^*(u^2) = \tfrac{1}{2}(-u^2 - u^2 i\sqrt{3}), \quad \theta^*(ui\sqrt{3}) = \tfrac{1}{2}(-ui\sqrt{3} - 3u),$$

$$\theta^*(u^2 i\sqrt{3}) = \tfrac{1}{2}(-u^2 i\sqrt{3} + 3u^2),$$

and so the required extension is defined by

$$\theta^*(a_1 + a_2 u + a_3 u^2 + a_4 i\sqrt{3} + a_5 ui\sqrt{3} + a_6 u^2 i\sqrt{3})$$
$$= \frac{1}{2}\Big(2a_1 + (-a_2 - 3a_5)u + (-a_3 + 3a_6)u^2 + 2a_4 i\sqrt{3}$$
$$+ (a_2 - a_5)ui\sqrt{3} + (-a_3 - a_6)u^2 i\sqrt{3}\Big).$$

It is possible, but unacceptably tedious, to verify directly that θ^* is a $\mathbb{Q}$-automorphism.

It seems clear, in the words of "1066 and All That", that normal extensions are a Good Thing. So it will be helpful to know that we can always extend a finite extension to make it normal. More precisely, if L is a finite extension of a field K, a field N containing L is said to be a **normal closure of L over K** if

(i) it is a normal extension of K; and

(ii) if E is a proper subfield of N containing L, then E is not a normal extension of K.

The following theorem states in effect that normal closures exist and are unique:

Theorem 7.18

Let L be a finite extension of a field K. Then,

(i) there exists a normal closure N of L over K;

(ii) if L' is a finite extension over K such that there is a K-isomorphism φ : $L \to L'$, and if N' is a normal closure of L' over K, then there is a K-isomorphism $\psi : N \to N'$ such that the diagram

$$
\begin{array}{ccccc}
K & \longrightarrow & L & \longrightarrow & N \\
\iota \downarrow & & \varphi \downarrow & & \psi \downarrow \\
K & \longrightarrow & L' & \longrightarrow & N'
\end{array}
$$

(in which ι is the identity map and unmarked maps are inclusions) commutes.

Proof

(i) Let $\{z_1, z_2, \ldots, z_n\}$ be a basis for L over K. Each z_i is algebraic over K, with minimum polynomial m_i (say). Let $m = m_1 m_2 \ldots m_n$, and let N be a splitting field for m over K. By the proof of Theorem 7.13, N is a normal extension of K. It contains all the roots of each of the polynomials m_i, and so certainly contains $z_1, z_2, \ldots, z_n$. Hence N contains L. Let E be a subfield of N containing L, and suppose that E is normal. For each i in $\{1, 2, \ldots, n\}$ the field E contains one root of m_i, namely z_i. By the definition of normality it follows that E contains all the roots of all the m_i, and so $E = N$. We have shown that N is a normal closure.

(ii) Let N' be a normal closure of L' over K. Every element of L has a unique expression $a_1 z_1 + a_2 z_2 + \cdots + a_n z_n$, where $a_1, a_2, \ldots, a_n \in K$. Let $u' = \varphi(u)$ be an arbitrary element of L'. Then there is a unique n-tuple $(a_1, a_2, \ldots, a_n)$ of elements of K such that

$$ u' = \varphi(u) = \varphi(a_1 z_1 + a_2 z_2 + \cdots + a_n z_n) = a_1 \varphi(z_1) + a_2 \varphi(z_2) + \cdots + a_n \varphi(z_n), $$

and it is easy to see that $\{\varphi(z_1), \varphi(z_2), \ldots, \varphi(z_n)\}$ is a basis for L' over K. The isomorphism φ also ensures that, for $i = 1, 2, \ldots, n$, the minimum polynomial of $\varphi(z_i)$ is $\hat{\varphi}(m_i)$ (where $\hat{\varphi}$ denotes the canonical extension of φ to the polynomial ring $L[X]$). Since N' is by assumption a normal extension of L', it must contain all the roots of all of the $\hat{\varphi}(m_i)$, and must in fact be a splitting field of $\hat{\varphi}(m) = \hat{\varphi}(m_1)\hat{\varphi}(m_2) \ldots \hat{\varphi}(m_n)$. The existence of the isomorphism ψ now follows from Theorem 5.3. $\square$

Corollary 7.19

Let L be a finite extension of K and let N be a normal closure of L. Then

$$ N = L_1 \vee L_2 \vee \cdots \vee L_k, $$

where $L_1, L_2, \ldots L_k$ are subfields containing K, each of them isomorphic to L.

Proof

By the theorem just proved, we may suppose that $L = K(z_1, z_2, \ldots, z_n)$, that $m_1, m_2, \ldots, m_n$ are (respectively) the minimum polynomials of $z_1, z_2, \ldots, z_n$, and that N is a splitting field over K for the polynomial $m_1 m_2 \ldots m_n$. Let $i \in \{1, 2, \ldots, n\}$ and let z_i' be a root of m_i. Then, for all choices of i and z_i', the field

$$K(z_1, \ldots, z_{i'}, \ldots z_n) \qquad (7.30)$$

is isomorphic to L. The field N is generated over K by the set $\{\alpha_1, \alpha_2, \ldots, \alpha_k\}$ of all the roots of all the polynomials $m_1, m_2, \ldots, m_n$, and hence by the fields of type (7.30). $\qquad \square$

Example 7.20

Determine the normal closure of $K = \mathbb{Q}(\sqrt[3]{2})$ over $\mathbb{Q}$.

Solution

A basis for K over $\mathbb{Q}$ is $\{1, u, u^2\}$, where $u = \sqrt[3]{2}$. The elements of the basis have minimum polynomials $X - 1$, $X^3 - 2$, $X^3 - 4$, respectively, and the routine in part (i) of the proof above would require us to find the splitting field of $(X - 1)(X^3 - 2)(X^3 - 4)$. Obviously the factor $X - 1$ is irrelevant here, since it already splits over $\mathbb{Q}$. We know that, over the field $\mathbb{Q}(u, i\sqrt{3})$,

$$X^3 - 2 = (X - u)(X - ue^{2\pi i/3})(X - ue^{-2\pi i/3}),$$

and it is easy to see that, over the same field,

$$X^3 - 4 = (X - u^2)(X - u^2 e^{2\pi i/3})(X - u^2 e^{-2\pi i/3}).$$

The conclusion is that the normal closure is $\mathbb{Q}(u, i\sqrt{3})$. $\qquad \square$

The following characterisation of normal extensions will be used later in the chapter:

Theorem 7.21

Let L be a finite normal extension of a field K, and let E be a subfield of L containing K. Then E is a normal extension of K if and only if every K-monomorphism of E into L is a K-automorphism of E.

Proof

Suppose first that E is a normal extension, so that E is its own normal closure. Let φ be a K-monomorphism from E into L, and let $z \in E$. Let $m = X^n + a_{n-1}X^{n-1} + \cdots + a_1X + a_0$ be the minimum polynomial of z over K. Then

$$z^n + a_{n-1}z^{n-1} + \cdots + a_1 z + a_0 = 0$$

and so, applying φ to this equality, we obtain

$$\big(\varphi(z)\big)^n + a_{n-1}\big(\varphi(z)\big)^{n-1} + \cdots + a_1\varphi(z) + a_0 = 0\,.$$

Thus $\varphi(z)$ is also a root of m in L. But z, an element of E, is a root of the irreducible polynomial m, and so, since E is normal, m splits completely over E. It follows that $\varphi(z) \in E$. Thus $\varphi(E)$ is a field contained in E. From Exercise 3.1,

$$[\varphi(E) : K] = [\varphi(E) : \varphi(K)] = [E : K] = [E : \varphi(E)][\varphi(E) : K]\,,$$

and so $\varphi(E) = E$. Thus φ is a K-automorphism of E.

Conversely, suppose that every K-monomorphism from E into L is a K-automorphism of E. Let f be an irreducible polynomial in $K[X]$ having a root z in E. To establish that E is a normal extension of K we require to show that f splits completely over E. Certainly, since L is normal, f splits completely over L. Let z' be another root of f in L. Then, by Corollary 7.16, there is a K-automorphism ψ of L such that $\psi(z) = z'$. Let ψ^* be the restriction of ψ to E. Then ψ^* is a K-monomorphism from E into L, and so, by our assumption, is a K-automorphism of E. Thus $z' = \psi(z) = \psi^*(z) \in E$, and we have shown that E is normal. $\qquad\square$

EXERCISES

7.7. Let L be a normal extension of a field K, and let E be a subfield of L containing K. Show that L is a normal extension of E.

7.8. Determine the normal closure of $\mathbb{Q}(\sqrt[4]{2})$ over $\mathbb{Q}$.

7.4 Separable Extensions

Some of the ideas in this section have already been touched upon in the last chapter, but it is useful at this stage to explore the topic a little further. If f is an irreducible polynomial with coefficients in a field K, the automorphisms in

Gal(f) permute the roots of f in the splitting field L. Since the study of these permutations would be hampered if f had repeated roots in L, there is a good case for restricting to extensions where this does not happen. An irreducible polynomial f with coefficients in a field K is said to be **separable over** K if it has no repeated roots in a splitting field. That is, in a splitting field L of f,

$$f = k(X - \alpha_1)(X - \alpha_2)\ldots(X - \alpha_n),$$

where the roots $\alpha_1, \alpha_2, \ldots, \alpha_n$ are all distinct. More generally,

- an arbitrary polynomial g in $K[X]$ is called **separable over** K if all its irreducible factors are separable over K;

- an algebraic element in an extension L of K is called **separable over** K if its minimum polynomial is separable over K;

- an algebraic extension L of K is called **separable** if every α in L is separable over K;

- a field K is called **perfect** if every polynomial in $K[X]$ is separable over K.

Separability is the second property (after normality) that will ensure that the maps Φ and Γ are mutually inverse. Fortunately separability is in the most interesting cases guaranteed, for we shall see that all fields of characteristic zero and all finite fields are perfect.

From Theorem 6.1 we know that the irreducible polynomial f has repeated roots in its splitting field if and only if f and Df have a non-trivial common factor. This is the key to the next observation.

Theorem 7.22

Let f be an irreducible polynomial with coefficients in a field K.

(i) If K has characteristic 0, then f is separable over K.

(ii) If K has finite characteristic p, then f is separable unless it is of the form

$$b_0 + b_1 X^p + b_2 X^{2p} + \cdots + b_m X^{mp}.$$

Proof

Let $f = a_0 + a_1 X + \ldots + a_n X^n$, with $\partial f = n \geq 1$, and suppose that f is not separable. Then f and Df have a common factor d of degree at least 1. Since f is irreducible, the factor d must be a constant multiple (an associate) of f, and this cannot divide Df unless

$$Df = a_1 + 2a_2 X + \cdots + na_n X^{n-1}$$

is the zero polynomial. Hence,

$$a_1 = 2a_2 = \cdots = na_n = 0. \tag{7.31}$$

If K has characteristic 0, this implies that f is the constant polynomial a_0, and we have a contradiction. Thus f must be separable.

Suppose now that char $K = p$. Then $ra_r = 0$ implies that $a_r = 0$ if and only if $p \nmid r$. Hence the only non-zero terms in f are of the form $a_{kp} X^{kp}$, for $k = 0, 1, 2, \ldots$. Writing a_{kp} as b_k gives the required conclusion. $\square$

From Part (i) of the theorem we immediately have the following conclusion:

Corollary 7.23

Every field of characteristic 0 is perfect.

For fields of finite characteristic the situation is more complicated. We must examine conditions under which a polynomial $f(X) = g(X^p) = b_0 + b_1 X^p + b_2 X^{2p} + \cdots + b_m X^{mp}$ is irreducible.

Theorem 7.24

Let K be a field with finite characteristic p, and let

$$f(X) = g(X^p) = b_0 + b_1 X^p + b_2 X^{2p} + \cdots + b_m X^{mp}.$$

Then the following statements are equivalent:

(i) f is irreducible in $K[X]$;

(ii) g is irreducible in $K[X]$, and not all of the coefficients b_i are pth powers of elements of K.

Proof

(i) $\Rightarrow$ (ii). If g has a non-trivial factorisation $g(X) = u(X)v(X)$, then f has a factorisation

$$f(X) = g(X^p) = u(X^p)v(X^p),$$

and we have a contradiction. Hence g is irreducible. If $b_i = c_i^p$ for $i = 1, 2, \ldots, m$, then, by Theorem 1.17,

$$f(X) = g(X^p) = c_0^p + (c_1 X)^p + \cdots + (c_m X^m)^p$$
$$= (c_0 + c_1 X + \cdots + c_m X_m)^p,$$

and again we have a contradiction. Hence not all of the coefficients b_i are pth powers.

(ii) $\Rightarrow$ (i). We shall in fact prove the (equivalent) contrapositive version, that $\neg$(i) $\Rightarrow$ $\neg$(ii). (Here the symbol $\neg$ stands for "not".) Suppose that f is reducible: we must prove either that g is reducible, or that all the coefficients of f are pth powers. We have two cases:

1. $f = u^r$, where $r > 1$ and u is irreducible;

2. $f = vw$, where $\partial v, \partial w > 0$, and v and w are coprime.

Case 1. Suppose first that $p \mid r$. Then $f = (u^{r/p})^p = h^p$ (say). If

$$h = d_0 + d_1 X + \cdots + d_s X^s,$$

then

$$f = h^p = (d_0 + d_1 X + \cdots + d_s X^s)^p = d_0^p + d_1^p X^p + \cdots + d_s^p X^{sp},$$

by Theorem 1.17, and so all the coefficients of f are pth powers. We have proved $\neg$(ii).

Next, suppose that $p \nmid r$. The definition of f in the statement of the theorem assures us that $Df = 0$; thus

$$0 = Df = r(Du)u^{r-1}$$

and so $Du = 0$. Thus we may write

$$u(X) = e_0 + e_1 X^p + \cdots + e_t X^{tp} = v(X^p),$$

and

$$g(X^p) = f(X) = \big(u(X)\big)^r = \big(v(X^p)\big)^r.$$

Thus $g(X) = \big(v(X)\big)^r$, and so g is not irreducible. Again, we have proved $\neg$(ii).

Case 2. Since $K[X]$ is a euclidean domain, there exist s, t in $K[X]$ such that

$$sv + tw = 1. \tag{7.32}$$

Also, from $Df = 0$ we deduce that

$$(Dv)w + v(Dw) = 0. \tag{7.33}$$

From (7.32) and (7.33) we have that

$$0 = (Dv)tw + tv(Dw) = (Dv)(1 - sv) + tv(Dw),$$

and so

$$Dv = sv(Dv) - tv(Dw).$$

Hence $v \mid Dv$. Since $\partial(Dv) < \partial v$, we must have that $Dv = 0$. Similarly, $Dw = 0$, and so we may write

$$v(X) = d_0 + d_1 X^p + \cdots + d_s X^{sp} \,,$$
$$w(X) = e_0 + e_1 X^p + \cdots + e_t X^{tp} \,.$$

If we define $\bar{v}(X) = d_0 + d_1 X + \cdots + d_s X^s$ and $\bar{w}(X) = e_0 + e_1 X + \cdots + e_t X^t$, then

$$g(X^p) = f(X) = v(X)w(X) = \bar{v}(X^p)\bar{w}(X^p) \,,$$

and so $g(X) = \bar{v}(X)\bar{w}(X)$. Thus g is not irreducible. Again, we have proved $\neg$(ii), and the proof is complete. $\square$

We can now establish the following result:

Theorem 7.25

Every finite field is perfect.

Proof

Let K be a finite field of characteristic p. Then (see Exercise 6.6) the Frobenius mapping $a \mapsto a^p$ is an automorphism of K, and so every element of K is a pth power. From Theorem 7.22, the only candidate for an inseparable irreducible polynomial is something of the form

$$f = b_0 + b_1 X^p + \cdots + b_m X^{mp} \,.$$

However, since all the coefficients are pth powers, Theorem 7.24 tells us that even polynomials of this form are reducible. Hence K is perfect. $\square$

Since all fields of characteristic zero and all finite fields are perfect, it is reasonable to ask whether there are any "imperfect" fields at all. Evidently, such a field has to be infinite and of finite characteristic, and so far we have not explicitly mentioned any such field. The most obvious example, however, is $K = \mathbb{Z}_p(X)$, the field of all rational forms with coefficients in $\mathbb{Z}_p$. For polynomials with coefficients in K we must use a different letter, such as Y, for the indeterminate. We look at the polynomial $Y^p - X$ in $K[Y]$. By Theorem 7.24, this is irreducible unless $-X$ is a pth power in the field K, that is, unless there exists an element $u(X)/v(X)$ in K such that $[u(X)/v(X)]^p = -X$. If we suppose that such an element exists, we deduce that $-X[v(X)]^p = [u(X)]^p$. But then $p \mid \partial\big([u(X)]^p\big)$ and $p \nmid \partial\big(X[v(X)]^p\big)$, and so we have a contradiction. Thus $f(Y) = Y^p - X$ is irreducible in $K[Y]$. Let L be a splitting field for f

over K, and let α be a root of f in L. Thus $\alpha^p = X$, and the factorisation of f in L is

$$f(Y) = Y^p - X = Y^p - \alpha^p = (Y - \alpha)^p \,.$$

The polynomial f is as inseparable as it is possible to be!

We shall have occasion later in the chapter to make use of the following observation:

Theorem 7.26

Let L be a finite separable extension of a field K, and let E be a subfield of L containing K. Then L is a separable extension of E.

Proof

Let $\alpha \in L$, and let m_K, m_E be the minimum polynomials of α over K and E, respectively. Suppose that m_K is separable. Within $E[X]$ we can use the division algorithm

$$m_K = qm_E + r \quad (\partial r < \partial m_E)\,,$$

and it follows that

$$r(\alpha) = m_K(\alpha) - q(\alpha)m_E(\alpha) = 0 - 0 = 0\,.$$

This is a contradiction to the minimality of the polynomial m_E unless $r = 0$. Hence $m_K = qm_E$ in the ring $E[X]$.

If m_E is not separable, then there is a non-constant polynomial g dividing m_E and Dm_E. Since $Dm_K = qDm_E + m_EDq$, it follows that g divides m_K and Dm_K. This can happen only if m_K has at least one repeated root in a splitting field, and so we have a contradiction. Hence m_E is separable. $\square$

Remark 7.27

We emphasise at this stage that, by Corollary 7.23, separability is guaranteed for fields of characteristic 0. In the next chapter, when we come to the applications of Galois theory to polynomial equations, we will (as is reasonable in a first course) confine ourselves to fields of characteristic zero, and so separability ceases to be an issue.

EXERCISES

7.9. The idea of a formal derivative can be extended to the field $K(X)$ of rational forms with coefficients in K by defining, for f, g ($\neq 0$) in

$K[X]$,
$$D(f/g) = (gDf - fDg)/g^2 \,.$$

Show that, for all u, v in $K(X)$,

$$D(u + v) = Du + Dv, \quad D(uv) = vDu + uDv \,,$$

$$D(u/v) = (vDu) - uDv)/v^2 \,.$$

7.10. Let K be a field with characteristic p. Show that K is perfect if and only if the Frobenius monomorphism $\varphi : a \mapsto a^p$ is an automorphism of K.

7.11. Let K be a field with characteristic p. An algebraic extension L of K is called **totally inseparable** if every element of $L \backslash K$ is inseparable. Show that every element of L has a minimum polynomial of the form $X^{p^n} + a_0$, where $a_0 \in K$.

7.5 The Galois Correspondence

A finite extension of a field K that is both normal and separable is called a **Galois extension**. The object of this section is to prove that for a Galois extension the mappings Γ and Φ are mutually inverse. This is a deep result, and we still have some spadework to do.

If we look at $\mathbb{Q}(\sqrt{2}, i\sqrt{3})$ and $\mathbb{Q}(\sqrt[3]{2}, i\sqrt{3})$, we notice that in both cases the order of the Galois group is equal to the degree over $\mathbb{Q}$ of the extension. Both of those examples are Galois extensions: they are certainly separable, by Corollary 7.23, and they are normal, being splitting fields (respectively) for $(X^2 - 2)(X^2 + 3)$ and $X^3 - 2$. We now set out to show that these are special cases of a general result. We shall prove first that, if $L : K$ is a normal, separable extension of degree n, and G is the Galois group of L over K, then $|G| = [L : K]$. In fact, it is useful to begin with something slightly more general:

Theorem 7.28

Let $L : K$ be a separable extension of finite degree n. Then there are precisely n distinct K-monomorphisms of L into a normal closure N of L over K.

Proof

The proof is by induction on the degree $[L : K]$. If $[L : K] = 1$, then $L = K = N$, and the only K-monomorphism of K into N is the identity mapping ι.

Suppose now that the result is established for all $n \leq k - 1$, and suppose that $[L : K] = k > 1$. Let $z_1 \in L \setminus K$, and let m (with $\partial m = r \geq 2$) be the minimum polynomial of z_1 over K. Thus $K \subset K(z_1) \subseteq L$, and $[K(z_1) : K] = r$. Then m, being irreducible and having one root z_1 in the normal extension N, splits completely over N. Since L is separable, the roots of m are all distinct: suppose that the roots are $z_1, z_2, \ldots, z_r$. Let $[L : K(z_1)] = s$; then $1 \leq s < k$, and $rs = k$.

The field N is a normal closure of L over $K(z_1)$, and so, by the induction hypothesis, we may suppose that the number of $K(z_1)$-monomorphisms from L into N is precisely s: denote them by $\mu_1, \mu_2, \ldots, \mu_s$. By Corollary 7.16 there are r distinct K-automorphisms $\lambda_1, \lambda_2, \ldots, \lambda_r$ of N, where $\lambda_i(z_1) = z_i$ ($i = 1, 2, \ldots, r$). Define maps $\varphi_{ij} : L \to N$ by

$$\varphi_{ij}(x) = \lambda_i\big(\mu_j(x)\big) \quad (x \in L; \; i = 1, 2, \ldots, r; \; j = 1, 2, \ldots, s). \qquad (7.34)$$

The definitions make it clear that the maps are all K-monomorphisms.

We show that the maps φ_{ij} are all distinct. First observe that

$$\varphi_{ij}(z_1) = \lambda_i\big(\mu_j(z_1)\big) = \lambda_i(z_1) = z_i. \qquad (7.35)$$

Hence, if $\varphi_{ij} = \varphi_{pq}$, it follows that $i = p$. Suppose now that $\varphi_{ij} = \varphi_{iq}$. Then, for all x in L,

$$\lambda_i\big(\mu_j(x)\big) = \lambda_i\big(\mu_q(x)\big).$$

Since λ_i is one–one, it follows that $\mu_j(x) = \mu_q(x)$ for all x in L, and so $j = q$. Thus the maps φ_{ij} are all distinct, and from (7.34) we now deduce that there are at least $rs = k$ distinct K-monomorphisms from L into N.

To show that there are no more than k, we must show that every K-monomorphism ψ from L into N coincides with one of the maps φ_{ij}. The map ψ must map z_1 to another root z_i of m in N. Let $\chi : L \to N$ be defined by

$$\chi(x) = \lambda_i^{-1}\big(\psi(x)\big).$$

This is certainly a K-monomorphism; indeed, since

$$\chi(z_1) = \lambda_i^{-1}\big(\psi(z_1)\big) = \lambda_i^{-1}(z_i) = z_1 \quad (x \in L),$$

it is a $K(z_1)$-monomorphism, and so must coincide with one of $\mu_1, \mu_2, \ldots, \mu_s$, say μ_j. Thus, for all x in L,

$$\mu_j(x) = \lambda_i^{-1}\big(\psi(x)\big),$$

and so $\psi(x) = \lambda_i\big(\mu_j(x)\big)$. Thus $\psi = \varphi_{ij}$. $\qquad \square$

If, in the statement of the Theorem 7.28, we suppose that L is normal as well as separable, then L is its own normal closure, and we obtain the following important corollary:

Corollary 7.29

Let L be a Galois extension of K, and let G be the Galois group of L over K. Then $|G| = [L : K]$.

We shall eventually see that normality and separability are the conditions required for the maps Γ and Φ defined by (7.11) and (7.12) to be mutually inverse. The next theorem establishes part of that result:

Theorem 7.30

Let L be a finite extension of K. Then $\Phi\big(\mathrm{Gal}(L : K)\big) = K$ if and only if L is a separable normal extension of K.

Proof

Suppose that L is a separable and normal extension of K, and let $[L : K] = n$. By Corollary 7.29, $|\mathrm{Gal}(L : K)| = n$. Denote $\Phi\big(\mathrm{Gal}(L : K)\big)$ by K'; then, from Theorem 7.11, we know that $K \subseteq K'$. By Theorem 7.12, we have that $[L : K'] = n$. Hence, since $K \subseteq K'$ and $[L : K] = [L : K']$, it follows from Exercise 3.1 that $K = K'$.

Conversely, suppose that $K = K'$. Let

$$\mathrm{Gal}(L : K) = \{\varphi_1 = \iota, \varphi_2, \ldots, \varphi_n\}.$$

Let f be an irreducible polynomial in $K[X]$ having a root z in L. To show that L is normal, we need to establish that f splits completely over L.

The images of z under the K-automorphisms $\varphi_1, \varphi_2, \ldots, \varphi_n$ need not all be distinct: we know that $\varphi_1(z) = z$, and we may re-label the elements of $\mathrm{Gal}(L : K)$ so that $\varphi_2(z), \ldots, \varphi_r(z)$ are the remaining distinct images of z under the automorphisms in $\mathrm{Gal}(L : K)$. For notational simplicity, let us write $\varphi_i(z) = z_i$ ($i = 1, 2, \ldots, r$). Note that $z_1 = z$.

Lemma 7.31

For each φ_j in $\mathrm{Gal}(L : K)$, the sets

$$\{z_1, z_2, \ldots, z_r\} \quad \text{and} \quad \{\varphi_j(z_1), \varphi_j(z_2), \ldots, \varphi_j(z_r)\}$$

are identical.

Proof

We note that $\varphi_j(z_i)$ is equal to $(\varphi_j\varphi_i)(z)$, and this is equal to z_k for some k, since $\varphi_j\varphi_i \in \mathrm{Gal}(L : K)$. Since φ_j is one–one, we conclude that it merely permutes the elements $z_1, z_2, \ldots, z_r$. $\square$

Now let g be the polynomial

$$(X - z_1)(X - z_2)\ldots(X - z_r) = X^r - e_1 X^{r-1} + \cdots + (-1)^r e_r, \qquad (7.36)$$

where the coefficients $e_1, e_2, \ldots, e_r$ are the elementary symmetric functions

$$e_1 = \sum_{i=1}^{r} z_i, \quad e_2 = \sum_{i \neq j} z_i z_j, \ldots, \quad e_r = z_1 z_2 \ldots z_r.$$

These coefficients are unchanged by any permutation of $z_1, z_2, \ldots, z_r$, and so, by Lemma 7.31, are unchanged by each φ_j in $\mathrm{Gal}(L : K)$. Thus g is a polynomial with coefficients in $\Phi\big(\mathrm{Gal}(L : K)\big)$, which (we are assuming) coincides with K.

Recall now that z was defined to be a root in L of the irreducible polynomial f in $K[X]$.

Lemma 7.32

The polynomial g defined by (7.36) is the minimum polynomial of z over K.

Proof

We must show that every polynomial in $K[X]$ having z as a root is divisible by g. So suppose that

$$h = a_0 + a_1 X + \cdots + a_m X^m,$$

with coefficients in K, is such that

$$a_0 + a_1 z + \cdots + a_m z^m = 0.$$

We can apply each φ_j to this relation: since φ_j leaves the coefficients a_i unchanged, we obtain

$$a_0 + a_1 z_j + \cdots + a_m z_j^m = 0 \quad (j = 1, 2, \ldots r),$$

and it follows that h is divisible by each of $X - z_1$, $X - z_2$, ..., $X - z_r$. Thus h is divisible by g. $\square$

Now, among the polynomials in $K[X]$ having a root z in L is the polynomial f with which (some time ago) we began. By Lemma 7.32, f is divisible by g, and so, since f was supposed to be irreducible, f is a constant multiple of g. Since g splits completely over L, so does f. Moreover, all its roots are distinct, and so L is, as required, a separable normal extension of K.   □

We end this section with another theorem concerning separable normal extensions:

Theorem 7.33

Let L be a Galois extension of a field K, and let E be a subfield of L containing K. If $\delta \in \mathrm{Gal}(L:K)$, then $\Gamma\big(\delta(E)\big) = \delta\Gamma(E)\delta^{-1}$.

Proof

Write $\delta(E) = E'$, $\Gamma(E) = H$ and $\Gamma(E') = H'$. We must show that $H' = \delta H\delta^{-1}$. Accordingly, let $\theta \in H$; we shall show that $\delta\theta\delta^{-1} \in H'$. Let $z' \in E'$ and let z be the unique element of E such that $\delta(z) = z'$. Then, since θ fixes all the elements of E,

$$(\delta\theta\delta^{-1})(z') = (\delta\theta\delta^{-1}\delta)(z) = \delta\big(\theta(z)\big) = \delta(z) = z',$$

and so $\delta\theta\delta^{-1} \in H'$. We have shown that $\delta H\delta^{-1} \subseteq H'$.

To show the opposite inclusion, let θ' be an arbitrary element of H', and let $z \in E$. Then $\delta(z) \in E'$, and so $\theta'\big(\delta(z)\big) = \delta(z)$. Hence

$$(\delta^{-1}\theta'\delta)(z) = (\delta^{-1}\delta)(z) = z,$$

and so $\delta^{-1}\theta'\delta \in \Gamma(E) = H$. We have shown that $\delta^{-1}H'\delta \subseteq H$, from which it follows immediately that $H' \subseteq \delta H\delta^{-1}$.   □

7.6 The Fundamental Theorem

This has been a long chapter. We finish it by gathering together all the bits and pieces in order to prove a theorem which, while easy to understand, has required a long sequence of preliminary results.

Theorem 7.34 (The Fundamental Theorem of Galois Theory)

Let L be a separable normal extension of a field K, with finite degree n.

(i) For all subfields E of L containing K, and for all subgroups H of the Galois group $\mathrm{Gal}(L : K)$,

$$\Phi\big(\Gamma(E)\big) = E , \quad \Gamma\big(\Phi(H)\big) = H .$$

Also,

$$|\Gamma(E)| = [L : E] \quad |\mathrm{Gal}(L : K)|/|\Gamma(E)| = [E : K] .$$

(ii) A subfield E is a normal extension of K if and only if $\Gamma(E)$ is a normal subgroup of $\mathrm{Gal}(L : K)$. If E is a normal extension, then $\mathrm{Gal}(E : K)$ is isomorphic to the quotient group $\mathrm{Gal}(L : K)/\Gamma(E)$.

Proof

(i) Let E be a subfield of L containing K. From Exercise 7.7 we know that L is a normal extension of E. Also, by Theorem 7.26, L is a separable extension of E. Hence, by Corollary 7.29, $|\Gamma(E)| = [L : E]$. From Theorem 3.3 and Corollary 7.29 it follows that

$$[E : K] = [L : K]/[L : E] = |\mathrm{Gal}(L : K)|/|\Gamma(E)| .$$

Since $\Gamma(E) = \mathrm{Gal}(L : E)$, it follows from Theorem 7.30 that

$$\Phi\big(\Gamma(E)\big) = E .$$

Now let H be any subgroup of the finite group $\mathrm{Gal}(L : K)$. From Theorem 7.11 we know that

$$H \subseteq \Gamma\big(\Phi(H)\big) . \tag{7.37}$$

Denote $\Gamma\big(\Phi(H)\big)$ by H'. From Exercise 7.3 we have that

$$\Phi(H) = \Phi\big(\Gamma[\Phi(H)]\big) = \Phi(H') .$$

From Theorem 7.12 we have that

$$|H| = [L : \Phi(H)] = [L : \Phi(H')] = |H'| .$$

This, together with (7.37) and the finiteness of $\mathrm{Gal}(L, K)$, tells us that $H' = H$. That is,

$$\Gamma\big(\Phi(H)\big) = H .$$

(ii) Suppose now that E is a normal extension. Let $\delta \in \mathrm{Gal}(L : K)$, and let δ' be the restriction of δ to E. Then δ' is a monomorphism from E into L and so, by Theorem 7.21, is a K-automorphism of E. Since $\delta(E) = \delta'(E) = E$, it follows by Theorem 7.33 that

$$\Gamma(E) = \Gamma\big(\delta(E)\big) = \delta\Gamma(E)\delta^{-1} .$$

Thus $\Gamma(E)$ is a normal subgroup of $\mathrm{Gal}(L:K)$.

Conversely, suppose that $\Gamma(E)$ is a normal subgroup of $\mathrm{Gal}(L:K)$. Let δ_1 be a K-monomorphism from E into L. By Corollary 7.14, this extends to a K-automorphism δ of L. The normality of $\Gamma(E)$ within $\mathrm{Gal}(L:K)$ means that $\delta\Gamma(E)\delta^{-1} = \Gamma(E)$, and hence, by Theorem 7.33,

$$\Gamma\big(\delta(E)\big) = \Gamma(E)\,.$$

Since Γ is one–one, it follows that $\delta(E) = \delta_1(E) = E$. Thus δ_1 is a K-automorphism of E. We have shown that every K-monomorphism of E into L is a K-automorphism of E. From Theorem 7.21 it follows that E is a normal extension of K.

It remains to show that, if E is a normal extension, then $\mathrm{Gal}(E:K) \simeq \mathrm{Gal}(L:K)/\Gamma(E)$. So suppose that E is normal and, as above, let δ' be the restriction to E of the K-automorphism δ of L. We have seen that $\delta' \in \mathrm{Gal}(E:K)$. Let $\Theta : \mathrm{Gal}(L:K) \to \mathrm{Gal}(E:K)$ be defined by

$$\Theta(\delta) = \delta'\,.$$

Then Θ is a group homomorphism: for all δ_1, δ_2 in $\mathrm{Gal}(L:K)$, with $\Theta(\delta_1) = \delta_1'$ and $\Theta(\delta_2) = \delta_2'$, and, for all z in E,

$$\begin{aligned}
\big([\Theta(\delta_1)][\Theta(\delta_2)]\big)(z) &= (\delta_1'\delta_2')(z) = \delta_1'\big(\delta_2(z)\big) \\
&= \delta_1\big(\delta_2(z)\big) = (\delta_1\delta_2)(z) \\
&= \big(\Theta(\delta_1\delta_2)\big)(z)\,.
\end{aligned}$$

Hence

$$[\Theta(\delta_1)][\Theta(\delta_2)] = \Theta(\delta_1\delta_2)\,.$$

The kernel of this homomorphism is the set of all δ in $\mathrm{Gal}(L:K)$ such that δ' is the identity map on E, and this is none other than $\Gamma(E)$. The result now follows from Theorem 1.20. $\qquad\qquad\square$

It is convenient at this point to establish two technical consequences of Theorem 7.34. First, let U and V be subgroups of a group G. Then it is a routine matter to show that $U \cap V$ is a subgroup of G. In general $U \cup V$ is not a subgroup, but there is always a smallest subgroup containing U and V, consisting of all products $u_1v_1u_2v_2 \ldots u_nv_n$ (for all n) with $u_1, u_2, \ldots \in U$, $v_1, v_2, \ldots \in V$. We denote this by $U \vee V$, and call it the **join** of U and V

Similarly, if E and F are subfields of a field K, then $E \cap F$ is also a subfield, and there is a subfield $E \vee F = E(F) = F(E)$, the **join** of E and F. The order-reversing Galois correspondence established in Theorem 7.34 has the following consequence:

Theorem 7.35

Let L be a Galois extension of finite degree over K, with Galois group G, and let E_1, E_2 be subfields of L containing K. If $\Gamma(E_1) = H_1$ and $\Gamma(E_2) = H_2$, then

$$\Gamma(E_1 \cap E_2) = H_1 \vee H_2, \quad \Gamma(E_1 \vee E_2) = H_1 \cap H_2.$$

Proof

Since $E_1 \subseteq E_1 \vee E_2$, it follows from the order-reversing property of the Galois correspondence that $\Gamma(E_1 \vee E_2) \subseteq \Gamma(E_1) = H_1$. Similarly, $\Gamma(E_1 \vee E_2) \subseteq H_2$, and so

$$\Gamma(E_1 \vee E_2) \subseteq H_1 \cap H_2.$$

To show the opposite inclusion, consider an element α of $H_1 \cap H_2$. Since $\alpha \in H_1 = \Gamma(E_1)$, $\alpha(x) = x$ for all x in E_1, and similarly $\alpha(y) = y$ for all y in E_2. Now, by Theorem 3.5, the elements of $E_1 \vee E_2 = E_1(E_2)$ are quotients of finite linear combinations (with coefficients in E_1) of finite products of elements of E_2, and so it follows that $\alpha(z) = z$ for all z in $E_1 \vee E_2$. Thus $\alpha \in \Gamma(E_1 \vee E_2)$, and so the first assertion of the theorem is proved.

From $E_1 \cap E_2 \subseteq E_1$ it follows that $H_1 = \Gamma(E_1) \subseteq \Gamma(E_1 \cap E_2)$. Similarly, $H_2 \subseteq \Gamma(E_1 \cap E_2)$, and so

$$H_1 \vee H_2 \subseteq \Gamma(E_1 \cap E_2).$$

To show the opposite inclusion, let x be an element of L not in $E_1 \cap E_2$ – say $x \notin E_1$. Since E_1 is precisely the fixed field of H_1, there exists γ in $H_1 \subseteq H_1 \vee H_2$ such that $\gamma(x) \neq x$. We deduce that $x \notin E_1 \cap E_2$ implies $x \notin \Phi(H_1 \vee H_2)$. That is, $\Phi(H_1 \vee H_2) \subseteq E_1 \cap E_2$, and the Galois correspondence gives $\Gamma(E_1 \cap E_2) \subseteq H_1 \vee H_2$. $\qquad\square$

In Chapter 8 we shall need the following theorem, due to Lagrange:

Theorem 7.36

Let K be a field of characteristic zero, and let $f \in K[X]$. Let

$$L = K(\alpha_1, \alpha_2, \ldots, \alpha_n)$$

be a splitting field for f over K. Let M be a field containing K, and let N be a splitting field of f over M. Then, up to isomorphism, L is a subfield of N, and $\mathrm{Gal}(N : M) \simeq \mathrm{Gal}(L : M \cap L)$.

Proof

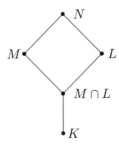

The field N is an extension of M, and hence of K, such that f splits completely in $N[X]$. Hence, by the definition of a splitting field, L is, up to isomorphism, a subfield of N, and we may write N as $M(\alpha_1, \alpha_2, \ldots, \alpha_n)$.

Let $H = \mathrm{Gal}(N : M)$, and let $\gamma \in H$. Then the restriction γ' of γ to L is a monomorphism from L into N. Since γ fixes the elements of M, it certainly fixes the elements of K; hence so does γ'. Moreover, since (by Theorem 7.9) γ maps each root α_i of f to another root of f, so also must γ'. The conclusion is that γ' is a monomorphism of L into itself. Since γ is an automorphism of $N = M(\alpha_1, \alpha_2, \ldots, \alpha_n)$, every root α_i of f is the image of some root of f under γ, and so also under γ'. Hence γ' maps onto $L = K(\alpha_1, \alpha_2, \ldots, \alpha_n)$ and so is a K-automorphism.

We thus have a mapping θ from H into the group $G = \mathrm{Gal}(L : K)$, given by $\theta(\gamma) = \gamma'$. The map is one–one, for if $\delta \in H$ and $\gamma' = \delta'$, then γ' and δ' act identically on the roots $\alpha_1, \alpha_2, \ldots \alpha_n$, and so $\gamma = \delta$. It is also a group homomorphism, since the restriction of $\gamma\delta$ to L is $\gamma'\delta'$. Thus $H \simeq \theta(H)$.

It remains to show that the image of θ is the subgroup $\mathrm{Gal}(L : M \cap L)$ of G. Since each γ in G fixes the elements of M, it is clear that each γ' fixes the elements of $M \cap L$. Thus $M \cap L \subseteq \Phi\big(\theta(H)\big)$, and so, by the Galois correspondence,

$$\theta(H) \subseteq \mathrm{Gal}(L : M \cap L). \tag{7.38}$$

Let x be an element of L not belonging to $M \cap L$. Thus $x \notin M$. Since M is the precise field whose elements are fixed by H, there is an element β in H for which $\beta(x) \neq x$. Then certainly $\big(\theta(\beta)\big)(x) \neq x$, and so $x \notin \Phi\big(\theta(H)\big)$. We have shown that $\Phi\big(\theta(H)\big) \subseteq M \cap L$, and it follows that

$$\mathrm{Gal}(L : M \cap L) \subseteq \theta(H). \tag{7.39}$$

From (7.38) and (7.39) we have that

$$\mathrm{Gal}(L : M \cap L) = \theta(H) \simeq H = \mathrm{Gal}(N : M).$$

$\square$

7.7 An Example

We round off this chapter with a fairly substantial example, one that illustrates most of the important features of the theory.

Example 7.37

Consider the Galois group $G = \text{Gal}[\mathbb{Q}(v,i) : \mathbb{Q}]$, where $v = \sqrt[4]{2}$. The field $\mathbb{Q}(v,i)$ is the splitting field of $X^4 - 2$ over $\mathbb{Q}$. If $\xi \in G$ then, by Theorem 7.9, $\xi(i) = \pm i$ and $\xi(v) \in \{v, iv, -v, -iv\}$. There are 8 elements in the group G:

$$
\begin{array}{llll}
\iota & : & v \mapsto v & \qquad \lambda & : & v \mapsto v \\
 & : & i \mapsto i & \qquad & : & i \mapsto -i \\[2mm]
\alpha & : & v \mapsto iv & \qquad \mu & : & v \mapsto iv \\
 & : & i \mapsto i & \qquad & : & i \mapsto -i \\[2mm]
\beta & : & v \mapsto -v & \qquad \nu & : & v \mapsto -v \\
 & : & i \mapsto i & \qquad & : & i \mapsto -i \\[2mm]
\gamma & : & v \mapsto -iv & \qquad \rho & : & v \mapsto -iv \\
 & : & i \mapsto i & \qquad & : & i \mapsto -i.
\end{array}
$$

The multiplication in G is given by the table as follows:

	ι	α	β	γ	λ	μ	ν	ρ
ι	ι	α	β	γ	λ	μ	ν	ρ
α	α	β	γ	ι	μ	ν	ρ	λ
β	β	γ	ι	α	ν	ρ	λ	μ
γ	γ	ι	α	β	ρ	λ	μ	ν
λ	λ	ρ	ν	μ	ι	γ	β	α
μ	μ	λ	ρ	ν	α	ι	γ	β
ν	ν	μ	λ	ρ	β	α	ι	γ
ρ	ρ	ν	μ	λ	γ	β	α	ι

This requires some computation: for example, from $\alpha\big(\lambda(v)\big) = \alpha(v) = iv$ and $\alpha\big(\lambda(i)\big) = -i$ we deduce that $\alpha\lambda = \mu$, and from $\lambda\big(\alpha(v)\big) = \lambda(iv) = \lambda(i)\lambda(v) = -iv$ and $\lambda\big(\alpha(i)\big) = -i$ we deduce that $\lambda\alpha = \rho$.

This group has three subgroups of order 4, namely,

$$H_1 = \{\iota, \alpha, \beta, \gamma\}, \quad H_2 = \{\iota, \beta, \lambda, \nu\}, \quad H_3 = \{\iota, \beta, \mu, \rho\}$$

and five subgroups of order 2, namely,

$$H_4 = \{\iota, \beta\}, \quad H_5 = \{\iota, \lambda\}, \quad H_6 = \{\iota, \mu\}, \quad H_7 = \{\iota, \nu\}, \quad H_8 = \{\iota, \rho\}.$$

It is easy to see that $\Phi(H_1) = \mathbb{Q}(i)$, and slightly less obvious that $\Phi(H_2) = \mathbb{Q}(v^2) = \mathbb{Q}(\sqrt{2})$ and $\Phi(H_3) = \mathbb{Q}(i\sqrt{2})$. Continuing, we find that

$$\Phi(H_4) = \mathbb{Q}(i, \sqrt{2}), \quad \Phi(H_5) = \mathbb{Q}(v), \quad \Phi(H_6) = \mathbb{Q}\big((1+i)v\big),$$

$$\Phi(H_7) = \mathbb{Q}(iv), \quad \Phi(H_8) = \mathbb{Q}\big((1-i)v\big).$$

The lattice of subgroups of G is

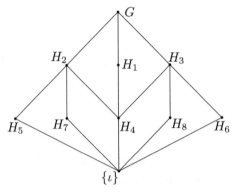

and the lattice of subfields E such that $\mathbb{Q} \subseteq E \subseteq \mathbb{Q}(v, i)$ is an upside down version of the same thing: (we write $\Phi(H_i)$ as F_i)

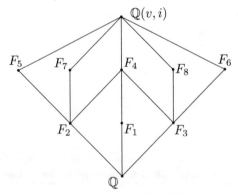

The normal subgroups of G are H_1, H_2, H_3 (of order 4), and H_4 (of order 2). The corresponding subfields $\mathbb{Q}(i)$, $\mathbb{Q}(\sqrt{2})$, $\mathbb{Q}(i\sqrt{2})$ and $\mathbb{Q}(i, \sqrt{2})$ are normal extensions, being the splitting fields (respectively) of $X^2 + 1$, $X^2 - 2$, $X^2 + 2$ and $(X^2 + 1)(X^2 - 2)$.

Remark 7.38

It will be important to note that $\mathrm{Gal}\big(\mathbb{Q}(v, i), \mathbb{Q}\big)$ is not abelian, although both $\mathrm{Gal}\big(\mathbb{Q}(v, i), \mathbb{Q}(i)\big) = \{\iota, \alpha, \beta, \gamma\}$ and

$$\mathrm{Gal}\big(\mathbb{Q}(i), \mathbb{Q}\big) \simeq \mathrm{Gal}\big(\mathbb{Q}(v, i), \mathbb{Q}\big)/\mathrm{Gal}\big(\mathbb{Q}(v, i), \mathbb{Q}(i)\big)$$

are abelian.

Remark 7.39

The example above is, of course, somewhat contrived – as indeed are the examples featuring in the exercises below, for it happens that we can easily factorise $X^4 - 2$ over the complex field. If we start with an irreducible polynomial such as

$$f = 2X^5 - 4X^4 + 8X^3 + 14X^2 + 7$$

(see Example 2.30) then it is by no means a trivial matter to determine the Galois group.

EXERCISES

7.12. Describe the Galois group $G = \mathrm{Gal}\big(\mathbb{Q}(u, i\sqrt{3}), \mathbb{Q}\big)$, where $u = \sqrt[3]{5}$. List the 4 proper subgroups of G, and describe the image under Φ of each of these subgroups.

7.13. Describe the Galois group $G = \mathrm{Gal}\big(\mathbb{Q}(\sqrt{2}, \sqrt{3}, \sqrt{5}), \mathbb{Q}\big)$. List the 14 proper subgroups of G, and describe the image under Φ of each of these subgroups.

8

Equations and Groups

8.1 Quadratics, Cubics and Quartics: Solution by Radicals

At this point we step back several centuries, indeed, in the case of quadratic equations, many centuries, for the procedure for solving quadratic equations can be traced back (see [2]) to the golden age of Babylon. Cubic and quartic equations were considered in the 16th and 17th centuries by Ferro[1], Tartaglia[2], Cardano[3], Ferrari[4] and Descartes[5].

It is clear that the roots of a polynomial equation

$$X^n + a_{n-1}X^{n-1} + \cdots + a_1 X + a_0 = 0$$

with rational coefficients are functions of those coefficients. All we are saying here is that the roots are determined by the coefficients, and it is legitimate to ask what kinds of functions are involved. For the linear equation $X + a_0$, the unique solution $-a_0$ is a *rational* function of the coefficients. In the case of a quadratic equation we can be quite explicit about the type of functions involved: the roots of

$$X^2 + a_1 X + a_0 \qquad (8.1)$$

---

[1] Scipione del Ferro, 1465–1526.
[2] Nicolo Tartaglia, 1499–1557.
[3] Girolamo Cardano, 1501–1576.
[4] Lodovico Ferrari, 1522–1565.
[5] René Descartes, 1596–1650.

are

$$\alpha = \tfrac{1}{2}(-a_1 + \sqrt{\Delta}), \quad \beta = \tfrac{1}{2}(-a_1 - \sqrt{\Delta}),$$

where $\Delta = a_1^2 - 4a_0$. The number Δ is referred to as the **discriminant** of the equation. The roots in general belong not to $\mathbb{Q}$, but to the extension $\mathbb{Q}(\sqrt{\Delta})$.

Before we leave quadratic equations, it is worth reminding ourselves that the sum and product of the roots of the equation (8.1) are given by

$$\alpha + \beta = -a_1, \quad \alpha\beta = a_0. \tag{8.2}$$

The cubic equation

$$X^3 + a_2 X^2 + a_1 X + a_0 = 0$$

requires a more substantial argument. First, if we make the substitution $X = Y - \tfrac{1}{3}a_2$, we obtain

$$Y^3 - a_2 Y^2 + \tfrac{1}{3}a_2^2 Y - \tfrac{1}{27}a_2^3 + a_2 Y^2 - \tfrac{2}{3}a_2^2 Y + \tfrac{1}{9}a_2^3 + a_1 Y - \tfrac{1}{3}a_1 a_2 + a_0 = 0,$$

which we can rewrite as

$$Y^3 + aY + b = 0.$$

We may thus confine our attention to cubic equations in which there is no quadratic term, and we can avoid some fractions if we write the standard cubic equation as

$$X^3 + 3aX + b = 0. \tag{8.3}$$

Let p be a root of the equation (8.3). We can certainly find q and r such that

$$q + r = p \text{ and } qr = -a : \tag{8.4}$$

by (8.2) they are the roots of the quadratic equation $X^2 - pX - a = 0$ (and will in general be complex numbers). Then

$$(q + r)^3 = q^3 + r^3 + 3(q^2 r + qr^2) = q^3 + r^3 + 3pqr$$

and so, by (8.4),

$$0 = p^3 + 3ap + b = q^3 + r^3 + 3p(a + qr) + b = q^3 + r^3 + b.$$

From $q^3 + r^3 = -b$ and $q^3 r^3 = -a^3$ we deduce from (8.2) that q^3 and r^3 are the roots of the equation

$$Z^2 + bZ - a^3 = 0.$$

Hence we may write

$$q^3 = \tfrac{1}{2}(-b + \sqrt{\Delta}), \quad r^3 = \tfrac{1}{2}(-b - \sqrt{\Delta}),$$

where $\Delta = b^2 + 4a^3$.

We find q and r, and hence p, by taking cube roots. More precisely, let q_1 and r_1 be cube roots (respectively) of q^3 and r^3, such that $q_1 r_1 = -a$. Then, if

$$\omega = e^{2\pi i/3} \quad \text{and} \quad \omega^2 = e^{4\pi i/3}$$

are the complex cube roots of unity, we also have

$$(q_1 \omega)(r_1 \omega^2) = -a \quad \text{and} \quad (q_1 \omega^2)(r_1 \omega) = -a .$$

Hence we have three possible values for p:

$$q_1 + r_1 , \quad q_1 \omega + r_1 \omega^2 , \quad q_1 \omega^2 + r_1 \omega ,$$

where

$$q_1 = \left[\tfrac{1}{2}\left(-b + \sqrt{b^2 + 4a^3}\right)\right]^{1/3} , \quad r_1 = \left[\tfrac{1}{2}\left(-b - \sqrt{b^2 + 4a^3}\right)\right]^{1/3} . \qquad (8.5)$$

Example 8.1

Find the three roots of
$$X^3 + 6X + 2 = 0 .$$

Solution

Here $a = b = 2$, and so $\Delta = b^2 + 4a^3 = 36$. It follows from (8.5) that $q_1 = 2^{1/3}$ and $r_1 = -4^{1/3} = -2^{2/3}$. (Note that $q_1 r_1 = -2$.) The three solutions are

$$q_1 + r_1 , \quad q_1 \omega + r_1 \omega^2 , \quad q_1 \omega^2 + r_1 \omega .$$

$\square$

That example, in which the discriminant has a rational square root, is perhaps a little contrived, for the discriminant may well be a complex number. Here is a more typical example, which looks on the surface very similar.

Example 8.2

Find the three roots of
$$X^3 - 6X + 2 = 0 .$$

Solution

Here $a = -2$ and $b = 2$, and so $\Delta = -28$. Thus $\sqrt{\Delta} = i\sqrt{28} = 2i\sqrt{7}$, and

$$q^3 = \tfrac{1}{2}(-2 + 2i\sqrt{7}) = -1 + i\sqrt{7} = \sqrt{8}e^{i\theta} ,$$

where $\cos\theta = -1/\sqrt{8}$, $\sin\theta = \sqrt{7}/\sqrt{8}$. Similarly, we see that $r^3 = \sqrt{8}e^{-i\theta}$. It follows that

$$q + r = \sqrt{2}(e^{i\theta/3} + e^{-i\theta/3}) = 2\sqrt{2}\cos(\theta/3)$$

is one of the roots of the equation. The other roots are $q\omega + r\omega^2$ and $q\omega^2 + r\omega$.

It is not hard to check that $q + r$ is a root: from the formula $\cos 3A = 4\cos^3 A - 3\cos A$, we have that

$$\begin{aligned}
(q+r)^3 - 6(q+r) + 2 &= 16\sqrt{2}\cos^3(\theta/3) - 12\sqrt{2}\cos(\theta/3) + 2 \\
&= 4\sqrt{2}\big(4\cos^3(\theta/3) - 3\cos(\theta/3)\big) + 2 \\
&= 4\sqrt{2}\cos\theta + 2 = 0.
\end{aligned}$$

$\square$

Example 8.3

Find the three roots of $X^3 - 3X + 2 = 0$.

Solution

This one is a bit silly, for the eagle-eyed reader may well have noticed that one of the roots is 1, and may even have noticed by differentiating the polynomial that 1 is a double root. It is perhaps of interest, however, to see what happens if we solve it using the general procedure. Here $a = -1$ and $b = 2$, and so $\Delta = 0$. Thus $q^3 = r^3 = -1$ and so, if we take $q = r = -1$, we obtain -2 as one of the roots. The others are $q\omega + r\omega^2 = -\omega - \omega^2 = 1$ and $q\omega^2 + r\omega = -\omega^2 - \omega = 1$ (since ω is a root of $X^2 + X + 1$). $\square$

The fundamental point to notice in the procedure for solving the cubic is that it is what is called a **solution by radicals**, by which we mean that the function

$$(a, b) \mapsto \left[\tfrac{1}{2}\big(-b + \sqrt{b^2 + 4a^3}\big)\right]^{1/3} + \left[\tfrac{1}{2}\big(-b - \sqrt{b^2 + 4a^3}\big)\right]^{1/3}$$

from the coefficients to the solution involves, in addition to rational operations, only the taking of square roots and cube roots.

We turn now to the quartic equation

$$X^4 + a_3 X^3 + a_2 X^2 + a_1 X + a_0 = 0,$$

where again a simple substitution means that we may consider only equations

$$X^4 + aX^2 + bX + c = 0$$

in which the cubic term is absent. Suppose that, over some extension of $\mathbb{Q}$, the polynomial factorises into quadratic factors:

$$X^4 + aX^2 + bX + c = (X^2 + pX + q)(X^2 - pX + r).$$

(The absence of a cubic term is reflected in the equal and opposite coefficients of X in the factors.) Then, equating coefficients, we see that

$$q + r - p^2 = a$$
$$p(r - q) = b$$
$$qr = c.$$

From the first two equations we see that

$$2pq = p^3 + ap - b, \quad 2pr = p^3 + ap + b, \tag{8.6}$$

and so, from the third equation

$$4p^2 c = (p^3 + ap - b)(p^3 + ap + b) = (p^3 + ap)^2 - b^2$$
$$= p^6 + 2ap^4 + a^2 p^2 - b^2.$$

Hence

$$p^6 + 2ap^4 + (a^2 - 4c)p^2 - b^2 = 0. \tag{8.7}$$

This is a cubic equation in p^2, and we may determine p^2 (and hence p) using the procedure of a cubic equation. Then, from (8.6) we determine q and r, and finally we solve the two quadratic equations

$$X^2 + pX + q = 0 \text{ and } X^2 - pX + r = 0. \tag{8.8}$$

Again this is a solution by radicals: the determination of p from (8.7) involves square and cube roots, the finding of q and r from (8.6) involves only rational operations, and the solving of the quadratic equations 8.8 involves square roots. It is certainly a cumbersome procedure, and when solving an equation with numerical coefficients it is almost invariably easier to use standard approximation procedures. But that is not the point here: we are investigating the *nature* of the solutions, not their numerical values.

These ingenious solutions gave the hope that equations of higher degree might yield to a similar approach, but no solution was found. The reason, it turned out, was simple: *there is no general procedure for solution by radicals for polynomials of degree greater than* 4. But to prove this we need first to clarify in field extension terms what we mean by a "solution by radicals", and then to develop some more group theory.

From this point on **we shall be dealing only with fields of characteristic** 0, which means (see Remark 7.27) that separability is not an issue.

Let K be a field. A field L containing K is called an **extension by radicals**, or a **radical extension**, if there is a sequence

$$K = L_0, L_1, \ldots, L_m = L$$

with the property that, for $j = 0, 1, \ldots, m-1$, $L_{j+1} = L_j(\alpha_j)$, where α_j is a root of an irreducible polynomial in $L_j[X]$ of the form $X^{n_j} - c_j$. This formalises the notion that the elements of L can be obtained from those of K by means of rational operations together with the taking of n_jth roots ($j = 1, 2, \ldots, m$): for example, if $K = \mathbb{Q}$, the element $(3 + \sqrt{2})^{1/7} + 5\sqrt[5]{2}(8 - \sqrt[3]{4})^{1/11}$ lies in a field L_5, where

$$
\begin{aligned}
L_1 &= \mathbb{Q}(\alpha_0)\,, & \alpha_0^2 &= 2 \in \mathbb{Q}\,, \\
L_2 &= L_1(\alpha_1)\,, & \alpha_1^7 &= 3 + \sqrt{2} \in L_1\,, \\
L_3 &= L_2(\alpha_2)\,, & \alpha_2^3 &= 4 \in L_2\,, \\
L_4 &= L_3(\alpha_3)\,, & \alpha_3^{11} &= 8 - \sqrt[3]{4} \in L_3\,, \\
L_5 &= L_4(\alpha_4)\,, & \alpha_4^5 &= 2 \in L_4\,.
\end{aligned}
$$

A polynomial f in $K[X]$ is said to be **soluble by radicals** if there is a splitting field for f contained in a radical extension of K. The conclusion of the ancient insights in Section 9.1 is that all linear, quadratic, cubic and quartic equations are soluble by radicals.

We shall need the following simple result later in the chapter:

Theorem 8.4

Let L be a radical extension of K, and let M be a normal closure of L. Then M is also a radical extension of K.

Proof

By Theorem 7.19, $M = L_1 \vee L_2 \vee \cdots \vee L_k$, where the extensions $L_1, L_2, \ldots, L_k$ are all isomorphic to L, and so all radical. The required result will follow if we prove that the join of two radical extensions is radical.

Let $M_1 = K(\alpha_1, \alpha_2, \ldots, \alpha_m)$, $M_2 = K(\beta_1, \beta_2, \ldots, \beta_n)$, where

$$
\begin{aligned}
\alpha_i^{k_i} &\in K(\alpha_1, \alpha_2, \ldots, \alpha_{i-1}) \ (i = 1, 2, \ldots m)\,, \\
\beta_j^{l_j} &\in K(\beta_1, \beta_2, \ldots, \beta_{j-1}) \ (j = 1, 2, \ldots n)\,.
\end{aligned}
$$

Then $M_1 \vee M_2 = K(\alpha_1, \alpha_2, \ldots, \alpha_m, \beta_1, \beta_2, \ldots, \beta_n)$, and

$$
\begin{aligned}
\alpha_i^{k_i} &\in K(\alpha_1, \alpha_2, \ldots, \alpha_{i-1}) \ (i = 1, 2, \ldots m)\,, \\
\beta_j^{l_j} &\in K(\alpha_1, \alpha_2, \ldots, \alpha_m, \beta_1, \beta_2, \ldots, \beta_{j-1}) \ (j = 1, 2, \ldots n)\,.
\end{aligned}
$$

Thus $M_1 \vee M_2$ is a radical extension. $\qquad\square$

EXERCISES

8.1. Find the roots of the equation $X^3 + 3X - 3 = 0$.

8.2. Find the roots of the equation $X^3 - 3X + 1 = 0$.

8.2 Cyclotomic Polynomials

Since a solution by radicals involves polynomials of the type $X^m - a$, it is appropriate that we should examine these more carefully than we have done thus far. We begin by looking at polynomials $f = X^m - 1$. We are confining ourselves to fields K of characteristic 0, and so can be sure that the splitting field L of f over K is both normal and separable. (If K has characteristic p and p divides m, then $Df = mX^{m-1} = 0$, and so (see Theorem 7.22) f is not separable.) The set R consisting of the roots in L of $X^m - 1$ is easily seen to be an (abelian) multiplicative subgroup of L. Indeed, we can be more precise:

Lemma 8.5

$(R, .)$ is a cyclic group.

Proof

Denote the exponent of R by e: thus $a^e = 1$ for all a in R. Since $X^e - 1$ has at most e roots, we must have $|R| \leq e$. However, the exponent of a group can never exceed the order of the group, and so $e \leq |R|$. Thus $e = |R| = m$ and so, by Corollary 6.5, R is cyclic. $\square$

Let ω be a **primitive** mth root of unity, namely, a generator of the cyclic group R. Then $R = \{1, \omega, \omega^2, \ldots, \omega^{m-1}\}$, and ω^j is a primitive mth root of unity if and only if j and m are coprime. Let P_m be the set of primitive mth root of unity. The **cyclotomic** polynomial Φ_m is defined by

$$\Phi_m = \prod_{\epsilon \in P_m} (X - \epsilon) \tag{8.9}$$

Some examples are helpful at this point.

Example 8.6

Let K be a field of characteristic 0, and let $L \subset \mathbb{C}$ be the splitting field for $X^p - 1$, where p is prime. Then, with the exception of the root 1, *all* of the roots of $X^p - 1$ are primitive, and so

$$\Phi_p = X^p + X^{p-1} + \cdots + X + 1.$$

Example 8.7

Let $K = \mathbb{Q}$ and let $L \subset \mathbb{C}$ be the splitting field of $X^{12} - 1$. One of the primitive 12th roots of unity is $\omega = e^{\pi i/6}$, and the elements of R are

$$1, \omega, \omega^2 = e^{\pi i/3}, \omega^3 = i, \omega^4 = e^{2\pi i/3}, \omega^5 = e^{5\pi i/6}, \omega^6 = -1,$$

$$\omega^7 = e^{7\pi i/6}, \omega^8 = e^{4\pi i/3}, \omega^9 = -i, \omega^{10} = e^{5\pi i/3}, \omega^{11} = e^{11\pi i/6}.$$

The group R contains the set P_d of primitive dth roots of unity, for each of the divisors $d = 12, 6, 4, 3, 2, 1$ of 12. Let

$$\Phi_d = \prod_{\epsilon \in P_d} (X - \epsilon).$$

The set P_{12} is

$$\{\omega, \omega^5, \omega^7 = \bar{\omega}^5, \omega^{11} = \bar{\omega}\},$$

and

$$\begin{aligned}
\Phi_{12} &= (X^2 - 2\cos\tfrac{\pi}{6} + 1)(X^2 - 2\cos\tfrac{5\pi}{6} + 1)\\
&= (X^2 - \sqrt{3}X + 1)(X^2 + \sqrt{3}X + 1)\\
&= X^4 - X^2 + 1.
\end{aligned}$$

The set P_6 is $\{\omega^2, \omega^{10} = \bar{\omega}^2\}$, and

$$\Phi_6 = X^2 - X + 1.$$

The set P_4 is $\{i, -i\}$, and

$$\Phi_4 = X^2 + 1.$$

The set P_3 is $\{\omega^4, \omega^8 = \bar{\omega}^4\}$, and

$$\Phi_3 = X^2 + X + 1.$$

The set P_2 is $\{\omega^6\}$, and $\Phi_2 = X + 1$. Finally, $P_1 = \{1\}$, and $\Phi_1 = X - 1$. Observe now that each Φ_d (where $d \mid 12$) is a polynomial with rational coefficients, and

$$X^{12} - 1 = \prod_{d \mid 12} \Phi_d = (X-1)(X+1)(X^2+X+1)(X^2+1)(X^2-X+1)(X^4-X^2+1).$$

This is no accident, as we shall see. Let K be a field of characteristic 0, and L a splitting field over K for $X^m - 1$. It is clear that, for all $m \geq 1$,

$$X^m - 1 = \prod_{d|m} \Phi_d$$

(where we are including both 1 and m among the divisors of m). What is less clear is that the polynomials Φ_d all lie in $K[X]$. The following lemma is the key:

Lemma 8.8

Let K, L be fields, with $K \subset L$. Let f, g be polynomials in $L[X]$ such that $f, fg \in K$. Then $g \in K$.

Proof

Let

$$f = a_0 + a_1 X + \cdots + a_m X^m, \quad g = b_0 + b_1 X + \cdots + b_n X^n,$$

where $a_0, a_1, \ldots, a_m \in K$, $b_0, b_1, \ldots, b_n \in L$, $a_m \neq 0$ and $b_n \neq 0$. Suppose that

$$fg = c_0 + c_1 X + \cdots + c_{m+n} X^{m+n} \in K[X].$$

Then $b_n = c_{m+n}/a_m \in K$. Suppose inductively that $b_j \in K$ for all $j > r$. Then

$$c_{m+r} = a_m b_r + a_{m-1} b_{r+1} + \cdots + a_{m-n+r} b_n,$$

where $a_i = 0$ if $i < 0$. Hence

$$b_r = (c_{m+r} - a_{m-1} b_{r+1} - \cdots - a_{m-n+r} b_n)/a_m \in K.$$

It follows that $b_j \in K$ for all j, and so $g \in K[X]$. $\square$

We can now easily prove the following result:

Theorem 8.9

Let K be a field of characteristic 0, containing mth roots of unity for each m, and let $K_0 (\simeq \mathbb{Q})$ be the prime subfield of K. Then, for every divisor d of m (including m itself), the cyclotomic polynomial Φ_d lies in $K_0[X]$.

Proof

It is clear that $\Phi_1 = X - 1$ belongs to $K_0[X]$. Let $d\ (\neq 1)$ be a divisor of m, and suppose inductively that $\Phi_r \in K_0[X]$ for all proper divisors r of d. Then, if Δ_d is the set of all divisors of d,

$$X^d - 1 = \left(\prod_{r \in \Delta_d \setminus \{d\}} \Phi_r \right) \Phi_d .$$

It follows from Lemma 8.8 that $\Phi_d \in K_0[X]$.   □

Remark 8.10

If $K = \mathbb{C}$ and $K_0 = \mathbb{Q}$, we can even assert that $\Phi_m \in \mathbb{Z}[X]$. See Exercise 8.4.

Example 8.11

By considering Φ_{14}, show that

$$\cos \frac{\pi}{7} + \cos \frac{3\pi}{7} + \cos \frac{5\pi}{7} = \frac{1}{2} .$$

Solution

Let $\omega = e^{\pi i/7}$; then the primitive roots of $X^{14} - 1$ are $\omega, \omega^3, \omega^5, \omega^9, \omega^{11}, \omega^{13}$, and so $\partial(\Phi_{14}) = 6$. Since $X^{14} - 1$ splits first as $(X^7 - 1)(X^7 + 1)$ and then into factors

$$X - 1,\ X + 1,\ X^6 + X^5 + X^4 + X^3 + X^2 + X + 1,\ X^6 - X^5 + X^4 - X^3 + X^2 - X + 1,$$

and since, by Example 8.6, the third factor in the list is Φ_7, we deduce that

$$\Phi_{14} = X^6 - X^5 + X^4 - X^3 + X^2 - X + 1. \tag{8.10}$$

The primitive roots are conjugate in pairs, and so Φ_{14} factorises in $\mathbb{R}[X]$ as

$$\left(X^2 - 2X \cos \frac{\pi}{7} + 1 \right)\left(X^2 - 2X \cos \frac{3\pi}{7} + 1 \right)\left(X^2 - 2X \cos \frac{5\pi}{7} + 1 \right). \tag{8.11}$$

Comparing the coefficients of X in (8.10) and (8.11) gives the required identity.
                                                                                                      □

We have already seen in Example 2.31 that Φ_m is irreducible (over $\mathbb{Q}$) if m is prime. In fact Φ_m is irreducible for every m, but the proof is surprisingly difficult:

Theorem 8.12

For all $m \geq 1$, the cyclotomic polynomial Φ_m is irreducible over $\mathbb{Q}$.

Proof

Suppose, for a contradiction, that Φ_m is not irreducible over $\mathbb{Q}$. From Exercise 8.4 below, we know that $\Phi_m \in \mathbb{Z}[X]$, and by Gauss's lemma (Theorem 2.24) we may suppose that $\Phi_m = fg$, where $f, g \in \mathbb{Z}[X]$ and f is an irreducible monic polynomial such that $1 \leq \partial f < \partial \Phi_m$.

Let K be a splitting field for Φ_m over $\mathbb{Q}$. At least one of the primitive mth roots of unity in K must be a root of f: let ϵ be one such. Since f is monic and irreducible and $f(\epsilon) = 0$, we may deduce that f is the minimum polynomial of ϵ over $\mathbb{Q}$. If p is a prime not dividing m, then ϵ^p is also a primitive mth root of unity. We show that ϵ^p is a root of f.

Suppose not. Then $g(\epsilon^p) = 0$. If we now define $h(X) \in \mathbb{Z}[X]$ by

$$h(X) = g(X^p),$$

it is clear that $h(\epsilon) = g(\epsilon^p) = 0$. We have already remarked that f is the minimum polynomial of ϵ over $\mathbb{Q}$, and so $f \mid h$: that is, $h = fu$, where $u \in \mathbb{Z}[X]$.

Consider now the map $n \mapsto \bar{n}$ from $\mathbb{Z}$ onto $\mathbb{Z}_p$, where $\bar{n}$ is the residue class $\{m \in \mathbb{Z} : m \equiv n \pmod{p}\}$. This map extends to a map $v \mapsto v^\dagger$ from $\mathbb{Z}[X]$ onto $\mathbb{Z}_p[X]$, in the obvious way:

$$(a_0 + a_1 X + \cdots + a_n X^n)^\dagger = \bar{a}_0 + \bar{a}_1 X + \cdots + \bar{a}_n X^n.$$

It is clear that $f^\dagger u^\dagger = h^\dagger$. On the other hand,

$$[h(X)]^\dagger = [g(X^p)]^\dagger = \left[(g(X))^\dagger\right]^p,$$

where the latter equality follows from repeated applications of the result that, in $\mathbb{Z}_p[X]$,

$$(ax + by)^p = a^p x^p + b^p y^p = ax^p + by^p.$$

Thus $f^\dagger u^\dagger = (g^\dagger)^p$.

Let $q^\dagger$ be an arbitrarily chosen irreducible factor of $f^\dagger$ in $\mathbb{Z}_p[X]$. Then $q^\dagger \mid (g^\dagger)^p$, and so $q^\dagger \mid g^\dagger$. Hence, since $q^\dagger$ divides both $f^\dagger$ and $g^\dagger$, we have that $(q^\dagger)^2 \mid \Phi_m^\dagger$. It follows that $\Phi_m^\dagger$, and hence also $X^m - 1$, has a repeated root in a splitting field over $\mathbb{Z}_p$. By Theorem 7.22, this cannot happen, since p does not divide m. Thus ϵ^p is a root of f.

Now let ζ be a root of f and η a root of g. Since both ζ and η are primitive mth roots of unity, we must have $\eta = \zeta^r$ for some r such that r and m are coprime. Let $r = p_1 p_2 \ldots p_k$, where $p_1, p_2, \ldots, p_k$ are (not necessarily distinct) primes not dividing m. From the conclusion of the last paragraph, we see that

$$\zeta^{p_1}, (\zeta^{p_1})^{p_2} = \zeta^{p_1 p_2}, \ldots, \zeta^{p_1 p_2 \cdots p_k} = \zeta^r$$

are all roots of f. Thus η is a root of f as well as g. It follows that η is a repeated root of Φ_m, and hence also of $X^m - 1$. From this contradiction we deduce that Φ_m is irreducible. $\qquad\square$

We now consider the Galois group of a polynomial $X^m - 1$:

Theorem 8.13

Let K be a field of characteristic zero, and let L be a splitting field over K of the polynomial $X^m - 1$. Then $\mathrm{Gal}(L : K)$ is isomorphic to R_m, the multiplicative group of residue classes $\bar{r}$ (mod m) such that $(r, m) = 1$.

Proof

Let ω be a primitive mth root of unity in L, and let $\sigma \in \mathrm{Gal}(L : K)$. Then $L = K(\omega)$. We know that $\sigma(\omega)$ must also be a primitive mth root of unity and so

$$\sigma \in \mathrm{Gal}(L : K) \text{ if and only if } \sigma(\omega) = \omega^{r_\sigma}, \text{ where } (r_\sigma, m) = 1. \qquad (8.12)$$

Since $\omega^r = \omega^s$ if and only if $r \equiv s$ (mod m), we have a one-one map from $\mathrm{Gal}(L : K)$ onto R_m, the multiplicative group of residue classes $\bar{r}$ mod m such that $(r, m) = 1$.

Let $\sigma, \tau \in \mathrm{Gal}(L : K)$. Then

$$(\sigma\tau)(\omega) = \sigma(\omega^{r_\tau}) = (\omega^{r_\tau})^{r_\sigma} = \omega^{r_\sigma r_\tau} = (\omega^{r_\sigma})^{r_\tau} = (\tau\sigma)(\omega), \qquad (8.13)$$

and so $\mathrm{Gal}(L : K)$ is abelian. The other consequence of (8.13) is that the map $\sigma \mapsto \bar{r}_\sigma$ is a homomorphism, since $\sigma\tau$ maps to $\bar{r}_\sigma \bar{r}_\tau$. It is clear that the map is one-one, and from (8.12) we deduce that it is also onto. $\qquad\square$

Corollary 8.14

Let K be a field of characteristic zero, and let L be a splitting field over K of the polynomial $X^p - 1$, where p is prime. Then $\mathrm{Gal}(L : K)$ is cyclic.

Proof

In the case where the exponent is prime, the Galois group is isomorphic to the multiplicative group $\mathbb{Z}_p^*$ of non-zero integers modulo p. By Theorem 6.3, this is a cyclic group. $\qquad\square$

Example 8.15

The splitting field in $\mathbb{C}$ of $X^8 - 1$ contains the primitive root $\omega = e^{\pi i/4}$. The Galois group has four elements, defined by

$$\omega \mapsto \omega,\ \omega \mapsto \omega^3,\ \omega \mapsto \omega^5,\ \omega \mapsto \omega^7,$$

and is isomorphic to $\{\bar{1}, \bar{3}, \bar{5}, \bar{7}\}$, with multiplication table

$\times$	$\bar{1}$	$\bar{3}$	$\bar{5}$	$\bar{7}$
$\bar{1}$	$\bar{1}$	$\bar{3}$	$\bar{5}$	$\bar{7}$
$\bar{3}$	$\bar{3}$	$\bar{1}$	$\bar{7}$	$\bar{5}$
$\bar{5}$	$\bar{5}$	$\bar{7}$	$\bar{1}$	$\bar{3}$
$\bar{7}$	$\bar{7}$	$\bar{5}$	$\bar{3}$	$\bar{1}$

EXERCISES

8.3. Let $p \neq 2$ be a prime. Show that

$$\Phi_{2p} = X^{p-1} - X^{p-2} + \cdots - X + 1.$$

8.4. Determine Φ_{15}.

8.5. Let P_m be the set of primitive mth roots of unity in the complex field $\mathbb{C}$. Show that the cyclotomic polynomial

$$\Phi_m = \prod_{\epsilon \in P_m} (X - \epsilon)$$

has integer coefficients.

8.6. Describe the Galois group of

$$\text{(i) } X^{12} - 1 \qquad \text{(ii) } X^{15} - 1.$$

8.3 Cyclic Extensions

Let char $K = 0$ and let $L : K$ be a field extension. We say that L is a **cyclic extension** of K if it is normal (and separable) and if $\mathrm{Gal}(L : K)$ is a cyclic group. Theorem 8.14 tells us in particular that, if p is prime, then the splitting field over K of $X^p - 1$ is a cyclic extension of K. We shall be interested in extensions whose Galois groups are "manageable", and in this section we investigate cyclic extensions more generally. Our goal will be to show that, under suitable conditions, cyclic extensions are radical extensions.

We begin with a result due to Hilbert[6]. To state the result we require some preliminaries. Let L be an extension, of finite degree n, of a field K (with char $K = 0$), and let N be a normal closure of L. By Theorem 7.28, there are exactly n distinct K-monomorphisms $\tau_1, \tau_2, \ldots \tau_n$ from L into N. For each element x of L, we define the **norm** $\mathrm{N}_{L/K}(x)$ and the **trace** $\mathrm{Tr}_{L/K}(x)$ by

$$\mathrm{N}_{L/K}(x) = \prod_{i=1}^{n} \tau_i(x), \quad \mathrm{Tr}_{L/K}(x) = \sum_{i=1}^{n} \tau_i(x). \tag{8.14}$$

Then we have

Theorem 8.16

The mapping $\mathrm{N}_{L/K}$ is a group homomorphism from $(L^*, .)$ into $(K^*, .)$. The mapping $\mathrm{Tr}_{L/K}$ is a non-zero group homomorphism from $(L, +)$ into $(K, +)$.

Proof

It is clear that, for all x, y in L^*,

$$\mathrm{N}_{L/K}(xy) = \prod_{i=1}^{n} \tau_i(xy) = \prod_{i=1}^{n} \tau_i(x)\tau_i(y)$$

$$= \Big(\prod_{i=1}^{n} \tau_i(x)\Big)\Big(\prod_{i=1}^{n} \tau_i(y)\Big) = \mathrm{N}_{L/K}(x)\mathrm{N}_{L/K}(y),$$

and similarly
$$\mathrm{Tr}_{L/K}(x + y) = \mathrm{Tr}_{L/K}(x) + \mathrm{Tr}_{L/K}(y);$$

thus $\mathrm{N}_{L/K}$ and $\mathrm{Tr}_{L/K}$ are (respectively) monomorphisms into $(L^*, .)$ and $(L, +)$. It remains to show that the images are contained in K.

Let τ be a K-automorphism of L. Then

$$\tau\tau_1, \tau\tau_2, \ldots, \tau\tau_n \tag{8.15}$$

---

[6] David Hilbert, 1862–1943.

are n distinct K-monomorphisms from L into N, and so the list (8.15) is simply the list $\tau_1, \tau_2, \ldots, \tau_n$ in a different order. Hence, for all x in L and all τ in $\mathrm{Gal}(L : K)$,

$$\tau\big(\mathrm{N}_{L/K}(x)\big) = \tau\Big(\prod_{i=1}^{n} \tau_i(x)\Big) = \prod_{i=1}^{n} \tau\big(\tau_i(x)\big)$$

$$= \prod_{i=1}^{n} \tau_i(x) \text{ (since multiplication is commutative)}$$

$$= \mathrm{N}_{L/K}(x),$$

and, similarly,

$$\tau\big(\mathrm{Tr}_{L/K}(x)\big) = \mathrm{Tr}_{L/K}(x).$$

Hence, by Theorem 7.34, both $\mathrm{N}_{L/K}(x)$ and $\mathrm{Tr}_{L/K}(x)$ lie in $\Phi\big(\mathrm{Gal}(L : K)\big) = K$.

It remains to show that $\mathrm{Tr}_{L/K}$ is not the zero homomorphism. Suppose, for a contradiction, that, for all x in L,

$$\mathrm{Tr}_{L/K}(x) = \tau_1(x) + \tau_2(x) + \cdots + \tau_n(x) = 0.$$

It follows that the set $\{\tau_1, \tau_2, \ldots, \tau_n\}$ is linearly dependent over L, and this contradicts Theorem 7.1. $\qquad\square$

We can now state Hilbert's theorem:

Theorem 8.17

Let L be a cyclic extension of a field K, and let τ be a generator of the (cyclic) group $\mathrm{Gal}(L : K)$. If $x \in L$, then $\mathrm{N}_{L/K}(x) = 1$ if and only if there is an element y in L such that $x = y/\tau(y)$, and $\mathrm{Tr}_{L/K}(x) = 0$ if and only if there is an element z in L such that $x = z - \tau(z)$.

Proof

Let $[L : K] = n$; then $\tau^n = \iota$, the identity automorphism. Suppose first that $x = y/\tau(y)$; then

$$\mathrm{N}_{L/K}(x) = \iota(x)\tau(x)\ldots\tau^{n-1}(x)$$

$$= \frac{y}{\tau(y)} \frac{\tau(y)}{\tau^2(y)} \frac{\tau^2(y)}{\tau^3(y)} \cdots \frac{\tau^{n-1}(y)}{\tau^n(y)}$$

$$= 1.$$

Conversely, suppose that $N_{L/K}(x) = 1$. Then

$$x^{-1} = \tau(x)\tau^2(x)\dots\tau^{n-1}(x)\,. \tag{8.16}$$

By Theorem 7.1, the set $\{\iota, \tau, \tau^2, \dots, \tau^{n-1}\}$ is linearly independent over L, and so the mapping

$$\iota + x\tau + x\tau(x)\tau^2 + \dots + x\tau(x)\tau^2(x)\dots\tau^{n-2}(x)\tau^{n-1}$$

is non-zero, which is to say that, for some t in L, the element

$$y = t + x\tau(t) + x\tau(x)\tau^2(t) + \dots + x\tau(x)\tau^2(x)\dots\tau^{n-2}(x)\tau^{n-1}(t)$$

is non-zero. Applying the automorphism τ gives

$$\tau(y) = \tau(t)+\tau(x)\tau^2(t)+\tau(x)\tau^2(x)\tau^3(t)+\dots+\tau(x)\tau^2(x)\tau^3(x)\dots\tau^{n-1}(x)\tau^n(t)\,. \tag{8.17}$$

Note also that

$$\begin{aligned}
x^{-1}y &= x^{-1}t + \tau(t) + \tau(x)\tau^2(t) + \tau(x)\tau^2(x)\tau^3(t) + \dots \\
&\quad \dots + \tau(x)\tau^2(x)\dots\tau^{n-2}(x)\tau^{n-1}(t) \\
&= \tau(t) + \tau(x)\tau^2(t) + \tau(x)\tau^2(x)\tau^3(t) + \dots \\
&\quad \dots + \tau(x)\tau^2(x)\dots\tau^{n-2}(x)\tau^{n-1}(t) + x^{-1}\tau^n(t)\,. \tag{8.18}
\end{aligned}$$

Comparing (8.17) and (8.18) and using (8.16) gives $\tau(y) = x^{-1}y$, and so $x = y/\tau(y)$, as required.

The closely similar proof concerning $\mathrm{Tr}_{L/K}$ is left as an exercise.

$\square$

Let K be a field of characteristic 0 and let $X^m - a \in K[X]$. Let L be a splitting field for $f = X^m - a$ over K. Then, by Theorem 7.22, f has distinct roots $\alpha_1, \alpha_2, \dots, \alpha_m$ in L, and so L contains the distinct roots

$$\alpha_1\alpha_1^{-1}, \alpha_2\alpha_1^{-1}, \dots, \alpha_m\alpha_1^{-1} \tag{8.19}$$

of the polynomial $X^m - 1$. Suppose, without loss of generality, that $\alpha_2\alpha_1^{-1} = \omega$ is a primitive mth root of unity. Then, in some order, the elements listed in (8.19) are the elements $1, \omega, \dots, \omega^{m-1}$, and so we can re-label the roots of $X^m - a$ in L as

$$\alpha_1, \omega\alpha_1, \dots, \omega^{m-1}\alpha_1\,. \tag{8.20}$$

Hence, over L,

$$X^m - a = (X - \alpha_1)(X - \omega\alpha_1)\dots(X - \omega^{m-1}\alpha_1)\,.$$

We have that

$$K \subseteq K(\omega) \subseteq L\,,$$

and the intermediate field $K(\omega)$ contains all the roots of unity.

We have established part of the following theorem:

Theorem 8.18

Let $f = X^m - a \in K[X]$, where K is a field of characteristic 0, and let L be a splitting field of f over K. Then L contains an element ω, a primitive mth root of unity. The group $\mathrm{Gal}(L : K(\omega))$ is cyclic, with order dividing m. The order is equal to m if and only if f is irreducible over $K(\omega)$.

Proof

We have seen that, if α is a root of f, then, over L,

$$f = (x - \alpha)(x - \omega\alpha)\ldots(X - \omega^{m-1}\alpha),$$

where ω is a primitive mth root of unity. Thus $L = K(\omega, \alpha)$, and an automorphism σ in $\mathrm{Gal}(L : K(\omega))$ is determined by its action on α. The image must be a root of f, and so

$$\sigma(\alpha) = \omega^{r_\sigma}\alpha$$

for some r_σ in $\{0, 1, \ldots, m - 1\}$. If τ is another element of $\mathrm{Gal}(L : K(\omega))$, then

$$(\sigma\tau)(\alpha) = \sigma(\omega^{r_\tau}\alpha) = \omega^{r_\tau}\omega^{r_\sigma}\alpha = \omega^{r_\tau + r_\sigma}\alpha,$$

and so $\sigma \mapsto \bar{r}_\sigma$ is a homomorphism onto the *additive* group $\mathbb{Z}_m$ of integers mod m. Moreover, $\bar{r}_\sigma = \bar{0}$ if and only if m divides r_σ, that is, if and only if $\sigma(\alpha) = \alpha$. The kernel of the homomorphism $\sigma \mapsto \bar{r}_\sigma$ is the identity in $\mathrm{Gal}(L : K(\omega))$, and so $\mathrm{Gal}(L : K(\omega))$ is isomorphic to a subgroup of the additive group $\mathbb{Z}_m$. From Exercises 1.27 and 1.28, we deduce that the group is cyclic.

Suppose now that $f = X^m - a$ is irreducible over $K(\omega)$. Then, by Corollary 7.29 and Theorem 3.7,

$$|\mathrm{Gal}(L : K(\omega))| = [L : K(\omega)] = \partial f = m,$$

and so $\mathrm{Gal}(L : K(\omega)) \simeq \mathbb{Z}_m$. Conversely, if f is not irreducible over $K(\omega)$, then it has a monic irreducible proper factor g such that $\partial g < m$. If ρ is a root of g in L, then

$$X^m - a = (X - \rho)(X - \omega\rho)\ldots(X - \omega^{m-1}\rho),$$

and so $L = K(\omega, \rho)$ is a splitting field for f over $K(\omega)$. Hence

$$|\mathrm{Gal}(L : K(\omega))| = [L : K(\omega)] = \partial g < m,$$

and so $\mathrm{Gal}(L : K(\omega))$ is isomorphic to a proper subgroup of $\mathbb{Z}_m$. $\qquad\square$

Remark 8.19

It is important to realise that, in the notation of the theorem just proved, although the Galois groups $\mathrm{Gal}\big(K(\omega):K\big)$ and $\mathrm{Gal}\big(L:K(\omega)\big)$ are both abelian, the group $\mathrm{Gal}(L:K)$ will usually be non-abelian. See Example 7.37, where we computed the Galois group of $X^4 - 2$, and the Remark 7.38.

The next result is a partial converse of Theorem 8.18:

Theorem 8.20

Let K be a field of characteristic zero, let m be a positive integer, and suppose that $X^m - 1$ splits completely over K. Let L be a cyclic extension of K such that $[L:K] = m$. Then there exists a in K such that $X^m - a$ is irreducible over K and L is a splitting field for $X^m - a$. Moreover, L is generated over K by a single root of $X^m - a$.

Proof

Here (in the notation of Theorem 8.18) $K(\omega) = K$. Let τ be a generator of the cyclic group $G = \mathrm{Gal}(L:K)$. Let ω be a primitive mth root of unity in K. Certainly every mth root of unity is left fixed by every automorphism in G. Hence $\mathrm{N}_{L/K}(\omega) = \omega^m = 1$. From Theorem 8.17 we deduce that there is an element z in L such that $\omega = z/\tau(z)$. Hence

$$\tau(z) = \omega^{-1} z, \tag{8.21}$$

and it easily follows that

$$\tau^k(z) = \omega^{-k} z \neq z \quad (k = 1, 2, \ldots, m - 1). \tag{8.22}$$

Thus $\Gamma[K(z)] = \{\iota\}$ and hence, since L is a cyclic extension (and so by definition normal) we may apply the Fundamental Theorem (Theorem 7.34) to obtain

$$K(z) = \Phi\big(\Gamma[K(z)]\big) = \Phi(\{\iota\}) = L.$$

From (8.21) we deduce that $\tau(z^m) = [\tau(z)]^m = \omega^{-m} z^m = z^m$, and it immediately follows that $\tau^k(z^m) = z^m$ for $k = 0, 1, \ldots, m - 1$. Thus $z^m \in \Phi(G) = K$. Denote z^m by a. Then z is a root of the polynomial $X^m - a$ in $K[X]$, and so the minimum polynomial g of z over K is a factor of $X^m - a$. Since $[K(z):K] = [L:K] = m$, the minimum polynomial g must be $X^m - a$. It follows that $X^m - a$ is irreducible over K. Moreover, the roots of $X^m - a$ are the elements $\omega^{-k} z$ ($k = 0, 1, \ldots m - 1$), all belonging to L, and so L is a splitting field for $X^m - a$ over K. $\square$

Less formally, the theorem just proved tells us that, provided the base field K has "enough" roots of unity, a cyclic extension of K is a radical extension.

In the proof of Theorem 8.18, the result depended on whether the polynomial $X^m - a$ was irreducible over $\mathbb{Q}(\omega)$. This can in practice be quite hard to determine, but if m is prime there is a useful result due to Abel:

Theorem 8.21 (Abel's theorem)

Let K be a field of characteristic 0, let p be a prime, and let $a \in K$. If $X^p - a$ is reducible over K, then it has a linear factor $X - c$ in $K[X]$.

Proof

Suppose that $f = X^p - a$ is reducible over K, and let g ($\in K[X]$) be a monic irreducible factor of f of degree d. If $d = 1$ there is nothing to prove; suppose that $1 < d < p$. Let L be a splitting field for f over K, and let β be a root of f in L. Then g factorises in $L[X]$ as

$$g = (X - \omega^{n_1}\beta)(X - \omega^{n_2}\beta)\dots(X - \omega^{n_d}\beta)\,, \tag{8.23}$$

where ω is a primitive pth root of unity and $0 \le n_1 < n_2 < \dots < n_d < p$. Suppose that

$$g = X^d - b_{d-1}X^{d-1} + \dots + (-1)^d b_0\,; \tag{8.24}$$

then, by comparing (8.23) and (8.24), we see that

$$b_0 = \omega^{n_1 + n_2 + \dots + n_d}\beta^d = \omega^n \beta^d \text{ (say)}\,.$$

Hence, since $\beta^p = a$,

$$b_0^p = \omega^{np}\beta^{dp} = \beta^{dp} = a^d\,.$$

Since p is prime, d and p have greatest common divisor 1, and so there exist integers s and t such that $sd + tp = 1$. Hence

$$a = a^{sd}a^{tp} = b_0^{sp}a^{tp} = (b_0^s a^t)^p$$

and so $X - c$, where $c = b_0^s a^t \in K$, is a linear factor of f. $\qquad\square$

Some examples at this stage are helpful:

Example 8.22

Determine the Galois group over $\mathbb{Q}$ of $X^5 - 7$.

Solution

By the Eisenstein criterion, the polynomial $X^5 - 7$ is irreducible over $\mathbb{Q}$. The primitive root $\omega = e^{2\pi i/5}$ has minimum polynomial $X^4 + X^3 + X^2 + X + 1$, and so $[\mathbb{Q}(\omega) : \mathbb{Q}] = 4$. The polynomial $X^5 - 7$ is even irreducible over $\mathbb{Q}(\omega)$. For if not, then by Abel's theorem there exists b in $\mathbb{Q}(\omega)$ such that $b = 7^{1/5}$. Since $[\mathbb{Q}(b) : \mathbb{Q}] \leq [\mathbb{Q}(\omega) : \mathbb{Q}] = 4$ and $[\mathbb{Q}(7^{1/5}) : \mathbb{Q}] \geq 5$, no such b can exist.

The roots of $X^5 - 7$ in $\mathbb{C}$ are

$$v, v\omega, v\omega^2, v\omega^3, v\omega^4,$$

where $v = 7^{1/5}$ and $\omega = e^{2\pi i/5}$. The Galois group consists of elements $\sigma_{p,q}$ ($p = 0, 1, 2, 3, 4$, $q = 1, 2, 3, 4$), where

$$\sigma_{p,q} : v \mapsto v\omega^p$$
$$: \omega \mapsto \omega^q.$$

The identity of the group is $\sigma_{0,1}$. Also,

$$\sigma_{p,q}\sigma_{r,s}(v) = \sigma_{p,q}(v\omega^r) = (v\omega^p)\omega^{qr} = v\omega^{p+qr},$$
$$\sigma_{p,q}\sigma_{r,s}(\omega) = = \sigma_{p,q}(\omega^s) = \omega^{qs},$$

and so

$$\sigma_{p,q}\sigma_{r,s} = \sigma_{p+qr,qs}, \tag{8.25}$$

where the addition and multiplication in the subscripts is mod 5. It is easy to show that, if $p \in \{1, 2, 3, 4, 5\}$ and $q \in \{1, 2, 3, 4\}$, then

$$\sigma_{1,1}^p = \sigma_{p,1}, \quad \sigma_{0,2}^q = \sigma_{0,2^q}, \quad \sigma_{p,1}\sigma_{0,2^q} = \sigma_{p,2^q};$$

hence the Galois group is generated by $\beta = \sigma_{1,1}$ and $\gamma = \sigma_{0,2}$, where

$$\beta^5 = 1, \ \gamma^4 = 1,$$

and

$$\gamma\beta = \sigma_{0,2}\sigma_{1,1} = \sigma_{2,2} = \beta^2\gamma.$$

The group, with presentation

$$\langle \beta, \gamma \mid \beta^5 = \gamma^4 = \beta^2\gamma\beta^{-1}\gamma^{-1} = 1 \rangle,$$

is of order 20. $\square$

EXERCISES

8.7. Let L be a cyclic extension of a field K, and let τ be a generator of the (cyclic) group $\text{Gal}(L : K)$.

(i) Show that, for each x in L, $\text{Tr}_{L/K}(x) = 0$ if and only if there exists an element z in L such that $x = z - \tau(z)$.

(ii) Show that $z - \tau(z) = z' - \tau(z')$ if and only if $z - z' \in K$.

8.8. Generalise Example 8.22 to the case of a polynomial $X^p - a$ in $K[X]$, where K has characteristic 0, p is prime, ω is a primitive pth root of unity and $X^p - a$ is irreducible over $K(\omega)$.

8.9. Let ω be a primitive 6th root of unity. Show that $X^6 - 3$ is not irreducible over $\mathbb{Q}(\omega)$. Describe the Galois group of $X^6 - 3$ over $\mathbb{Q}$.

9
Some Group Theory

Introduction

In this chapter we briefly stand aside from the main issue in order to examine the aspects of group theory that we shall need. Proofs are provided for the sake of completeness, but you may prefer simply to note the key results, which are Theorems 9.4, 9.6, 9.16, 9.19, 9.20, 9.23, 9.24 and 9.25.

9.1 Abelian Groups

It is traditional to write abelian groups in additive notation, writing $a + b$, 0, $-a$ and na (with $n \in \mathbb{Z}$) rather than ab, 1, a^{-1} and a^n. We shall be concerned here solely with *finite* abelian groups.

An abelian group A with subgroups $U_1, U_2, \ldots, U_k$ is said to be the **direct sum** of $U_1, U_2, \ldots, U_k$ if every element a of A has a unique expression

$$a = u_1 + u_2 + \cdots + u_k, \qquad (9.1)$$

where $u_i \in U_i$ ($i = 1, 2, \ldots k$). It follows that $U_i \cap U_j = \{0\}$ whenever $i \neq j$, for if w were a non-zero element in $U_i \cap U_j$, it would have two distinct expressions of the type (9.1), one in which $u_i = w$ and $u_j = 0$, the other in which $u_i = 0$ and $u_j = w$. We write

$$A = U_1 \oplus U_2 \oplus \cdots \oplus U_k.$$

One important immediate consequence of the definition is that, for all $u_i \in U_i$ $(i = 1, 2, \ldots, k)$,

$$u_1 + u_2 + \cdots + u_k = 0 \text{ implies } u_1 = u_2 = \cdots = u_k = 0; \qquad (9.2)$$

for otherwise we would have two distinct expressions for the element 0, the other being

$$0 + 0 + \cdots + 0.$$

In fact the condition (9.2) is *equivalent* to the uniqueness condition at (9.1); for if

$$a = u_1 + u_2 + \cdots + u_k = u'_1 + u'_2 + \cdots + u'_k$$

(with $u_i, u'_i \in U_i$ for all i), then

$$(u_1 - u'_1) + (u_2 - u'_2) + \cdots + (u_k - u'_k) = 0,$$

and it follows immediately from (9.2) that $u_i = u'_i$ for all i.

Lemma 9.1

Let a be an element of a finite abelian group A, and suppose that the order of a is mn, where $\gcd(m, n) = 1$. Then a can be written in exactly one way as $b + c$, where $o(b) = m$ and $o(c) = n$.

Proof

Let $b' = na$ and $c' = ma$. Then certainly $o(b') = m$ and $o(c') = n$. Since m and n are coprime, there exist s, t in $\mathbb{Z}$ such that $sm + tn = 1$. Hence $a = (sm + tn)a = tb' + sc'$. Certainly $\gcd(t, m) = 1$, since any non-trivial common divisor of t and m would have to divide $sm + tn$, and this cannot happen. Similarly $\gcd(s, n) = 1$. It follows that $o(tb') = m$ and $o(sc') = n$. Thus $b = tb'$ and $c = sc'$ are elements such that $a = b + c$.

To prove uniqueness, suppose that $a = b + c = b_1 + c_1$, where $o(b) = o(b_1) = m$ and $o(c) = o(c_1) = n$. Then $b - b_1 = c_1 - c = d$ (say). Then $md = mb - mb_1 = 0$, and $nd = nc_1 - nc = 0$, and so $o(d)$ divides both m and n. Hence $o(d) = 1$, and so $b - b_1 = c_1 - c = 0$. $\qquad \square$

It is easy to extend the argument above to obtain the following corollary:

Corollary 9.2

Let a be an element of a finite abelian group A, and suppose that $o(a) = m_1 m_2 \ldots m_r$, where $\gcd(m_i, m_j) = 1$ whenever $i \neq j$. Then a can be written in exactly one way as $a_1 + a_2 + \cdots + a_r$, where $o(a_i) = m_i$ $(i = 1, 2, \ldots, r)$.

Proof

Since $\gcd(m_1 \ldots m_{r-1}, m_r) = 1$, we can use the theorem to write a uniquely as $a' + a_r$, with $o(a') = m_1 \ldots m_{r-1}$ and $o(a_r) = m_r$. The result then follows by induction on r. $\qquad \square$

Suppose now that A is an abelian group of order

$$n = p_1^{e_1} p_2^{e_2} \ldots p_r^{e_r} \,.$$

Let U_i be the set of elements of A whose order is a power of p_i. Then U_i is a subgroup of A. For suppose that $x, y \in U_i$, with orders p_i^k, p_i^l, respectively; then

$$p_i^{\max\{k,l\}}(x - y) = 0 \,,$$

and so the order of $x - y$, being a divisor of $p_i^{\max\{k,l\}}$, is a power of p_i. Thus $x - y \in U_i$.

Let a be an element of A. Then a has order $p_1^{d_1} p_2^{d_2} \ldots p_r^{d_r}$ dividing n. By Corollary 9.2, a can be expressed uniquely as $a_1 + a_2 + \cdots + a_r$, with $o(a_i) = p_i^{d_i}$ $(i = 1, 2, \ldots, r)$. Thus

$$A = U_1 \oplus U_2 \oplus \cdots \oplus U_r \,.$$

We have proved

Theorem 9.3

Every finite abelian group is expressible as the direct sum of abelian p-groups.

This result is an important step on the way to establishing the basis theorem:

Theorem 9.4 (The Basis Theorem)

Every finite abelian group is expressible as a direct sum of cyclic groups.

Proof

In view of Theorem 9.3, we need only consider an abelian p-group A, of order p^m. Let a_1 be an element of maximal order p^{r_1} in A, and let $A_1 = \langle a_1 \rangle$, the cyclic subgroup of A generated by a_1. If $r_1 = m$, then $\langle a_1 \rangle = A$ and we have nothing to prove, for the group A is cyclic.

So suppose that $r_1 < m$. We prove the result by induction. Suppose that we have found k elements

$$a_1, a_2, \ldots, a_k$$

of orders

$$p^{r_1}, p^{r_2}, \ldots, p^{r_k}$$

(respectively) such that

(i) $r_1 \geq r_2 \geq \cdots \geq r_k$;

(ii) the subgroup $P_k = \langle a_1, a_2, \ldots, a_k \rangle$ is the direct sum

$$\langle a_1 \rangle \oplus \langle a_2 \rangle \oplus \cdots \langle a_k \rangle \, ;$$

(iii) no element of $A \setminus P_k$ has order exceeding p^{r_k}.

If $P_k = A$, then we are home. So suppose that there exists b in $A \setminus P_k$. By (iii), the order of b is p^β, where $\beta \leq r_k$. The set of multiples of b lying in P_k must be non-empty, since $p^\beta b = 0 \in P_k$: let λ be the *least* positive integer with the property that $\lambda b \in P_k$. Thus

$$\lambda b = \sum_{i=1}^{k} \mu_i a_i \quad (\lambda \leq p^\beta) \, . \tag{9.3}$$

The integer λ must in fact be a power of p. To see this we divide p^β by λ to obtain $p^\beta = q\lambda + r$, with $0 \leq r < \lambda$. If $r \neq 0$, then

$$rb = p^\beta b - q\lambda b = -q\lambda b \in P_k \, ,$$

contradicting the definition of λ as the *least* integer with this property. Hence $r = 0$ and so λ divides p^β. This can happen only if λ is a power of p: write $\lambda = p^{r_{k+1}}$. Certainly $r_{k+1} \leq r_k$ (by (iii)), and $r_{k+1} \leq \beta$.

We show next that every coefficient μ_i featuring in (9.3) is divisible by λ. Multiply (9.3) by $p^\beta / \lambda = p^{\beta - r_{k+1}}$ to obtain

$$0 = p^\beta b = \sum_{i=1}^{k} (\mu_i p^\beta / \lambda) a_i \, .$$

It follows from (ii) that $(\mu_i p^\beta / \lambda) a_i = 0$ for all i, and hence that $\mu_i p^\beta / \lambda = \mu_i p^{\beta - r_{k+1}}$ is divisible by $o(a_i) = p^{r_i}$: write $\mu_i p^\beta / \lambda = \mu_i' p^{r_i}$. Since $\beta \leq r_i$ for $i = 1, 2, \ldots, k$, we may rewrite this as

$$\mu_i = \lambda \mu_i' p^{r_i - \beta} = \lambda \nu_i \, , \tag{9.4}$$

where $\nu_i = \mu_i' p^{r_i - \beta}$ is an integer.

Let

$$a_{k+1} = b - \sum_{i=1}^{k} \nu_i a_i \,. \tag{9.5}$$

Then the order of a_{k+1} is $\lambda = p^{r_{k+1}}$. For, from (9.3) and (9.4),

$$\lambda a_{k+1} = \lambda b - \sum_{i=1}^{k} \lambda \nu_i a_i = 0 \,,$$

and, if $\kappa a_{k+1} = 0$ for $\kappa > 0$, then $\kappa b \in P_k$ and so $\kappa \geq \lambda$.

Let $P_{k+1} = \langle a_1, a_2, \ldots, a_k, a_{k+1} \rangle$. It remains to show that

$$P_{k+1} = \langle a_1 \rangle \oplus \langle a_2 \rangle \oplus \cdots \langle a_k \rangle \oplus \langle a_{k+1} \rangle \,.$$

We show that, if

$$z_1 a_1 + z_2 a_2 + \cdots + z_{k+1} a_{k+1} = 0 \,, \tag{9.6}$$

where $z_1, z_2, \ldots, z_{k+1}$ are integers, then $z_1 a_1 = z_2 a_2 = \cdots = z_{k+1} a_{k+1} = 0$. So suppose that (9.6) holds. Then $z_{k+1} a_{k+1}$ belongs to P_k and, by (9.5), so does $z_{k+1} b$. By the minimal property of λ, we deduce that $\lambda \leq z_{k+1}$. The division algorithm gives $z_{k+1} = q\lambda + r$, with $0 \leq r < \lambda$, and so $rb = z_{k+1} b - \lambda b \in P_k$, a contradiction unless $r = 0$. Thus λ *divides* z_{k+1}:

$$z_{k+1} = \lambda z'_{k+1} = p^{r_{k+1}} z'_{k+1} \,,$$

and it follows, since the order of a_{k+1} is $\lambda = p^{r_{k+1}}$, that $z_{k+1} a_{k+1} = 0$. It then immediately follows from (ii) that $z_i a_i = 0$ for $i = 1, 2, \ldots, k$, and so

$$P_{k+1} = \langle a_1, a_2, \ldots, a_{k+1} \rangle = \langle a_1 \rangle \oplus \langle a_2 \rangle \oplus \cdots \oplus \langle a_{k+1} \rangle \,.$$

Since A is finite, the process must eventually terminate, and we find that

$$A = \langle a_1, a_2, \ldots, a_l \rangle = \langle a_1 \rangle \oplus \langle a_2 \rangle \oplus \cdots \oplus \langle a_l \rangle \,,$$

a direct sum of cyclic groups.   □

While the additive notation for abelian groups is helpful, it is natural to use multiplicative notation for abelian Galois groups. The definition of a direct sum is easily rewritten in multiplicative notation, and we usually then prefer call it a **direct product**, and to write $U_1 \times U_2 \times \cdots \times U_k$. We have subgroups (necessarily normal since A is abelian)

$$\{1\} = V_0 \lhd V_1 \lhd \cdots \lhd V_k = A \,, \tag{9.7}$$

where $V_i = U_1 \times U_2 \times \cdots \times U_i \quad (i = 1, 2, \ldots, k)$.

Theorem 9.5

With the above notation, V_i/V_{i-1} is isomorphic to U_i.

Proof

Let $\varphi : V_i \to U_i$ be given by

$$\varphi(v_i) = u_i \,,$$

where $u_1 u_2 \ldots u_i$ is the unique expression of v_i as a product of elements from $U_1, U_2, \ldots, U_i$. It is clear that φ maps onto U_i. Also, φ is a homomorphism, for if $v_i' = u_1' u_2' \ldots u_i' \in V_i$, then

$$\varphi(v_i v_i') = \varphi[(u_1 u_1')(u_2 u_2') \ldots (u_i u_i')] = u_i u_i' = \varphi(v_i)\varphi(v_i') \,.$$

The kernel of φ is

$$\{u_1 u_2 \ldots u_i \ : \ u_i = 1\} = V_{i-1} \,,$$

and so, by Theorem 1.20, $U_i \simeq V_i/V_{i-1}$. $\qquad\qquad\square$

A finite group is called **soluble**[1] if, for some $m \geq 0$, it has a finite series

$$\{1\} = G_0 \subseteq G_1 \subseteq \cdots \subseteq G_m = G \tag{9.8}$$

of subgroups such that, for $i = 0, 1, \ldots, m - 1$,

(i) $G_i \lhd G_{i+1}$,

(ii) G_{i+1}/G_i is cyclic.

Note carefully that we are *not* saying that the subgroups G_i are all normal in G. (See Exercise 9.3.)

From (9.7) we immediately deduce:

Theorem 9.6

Every finite abelian group is soluble.

EXERCISES

9.1. Let H be a subgroup of a group G and let N be a normal subgroup of G such that $N \subseteq H$. Show that $N \lhd H$.

---

[1] The American term is *solvable*.

9.2. Let H be a subgroup of G and let N_1, N_2 be subgroups of G such that $N_1 \lhd N_2$. Show that $H \cap N_1 \lhd H \cap N_2$.

9.3. Give an example of a group G containing subgroups G_1, G_2 such that $G_1 \lhd G_2$ and $G_2 \lhd G$, but such that G_1 is not normal in G.

9.4. Show that there is an alternative definition of a finite soluble group, in which the quotients G_{i+1}/G_i are abelian rather than cyclic.

9.2 Sylow Subgroups

We begin with a general result in group theory, but before stating the result we need to make an observation about products of subgroups. If H and K are subgroups of a group G, then the subgroup $H \vee K$, the smallest subgroup of G containing H and K, consists of all finite products $y = h_1 k_1 h_2 k_2 \ldots h_m k_m$, where $h_1, h_2, \ldots, h_m \in H$ and $k_1, k_2, \ldots, k_m \in K$. If at least one of the subgroups, say H, is normal, then we can rewrite $k_1 h_2$ as $h'_2 k_1$, where $h'_2 = k_1 h_2 k_1^{-1} \in H$. By repeating this argument, we can obtain an expression $h^* k^*$ for y, and it is then natural to write $H \vee K$ as HK (or equivalently as KH).

Theorem 9.7

Let G be a group, let $N \lhd G$ and let H be a subgroup of G.

(i) $N \cap H \lhd H$ and $H/(N \cap H) \simeq NH/H$.

(ii) If $N \subseteq H$ and $H \lhd G$, then $N \lhd H$, $H/N \lhd G/N$, and $(G/N)/(H/N) \simeq G/N$.

Proof

(i) Let $x \in N \cap H$ and $h \in H$. Then $h^{-1} x h \in N \cap H$, and so $N \cap H \lhd H$. Let $\phi : g \mapsto Ng$ be the natural mapping from G onto G/N, and let $\iota : H \to G$ be the inclusion mapping. Then the image of the homomorphism $\phi \circ \iota : H \to G/N$ is NH/N, and the kernel is $N \cap H$ and so, by Theorem 1.20, $H/(N \cap H) \simeq NH/H$.

(ii) It is clear that $N \lhd H$ (see Exercise 9.1). Define a mapping $\theta : G/N \to G/H$ by the rule that

$$\theta(Ng) = Hg.$$

This is well defined: if $Ng_1 = Ng_2$, then $g_1 g_2^{-1} \in N \subseteq H$, and so $Hg_1 = Hg_2$. It clearly maps onto G/H. It is a homomorphism:

$$\theta\big((Na)(Nb)\big) = \theta\big(N(ab)\big) = H(ab) = (Ha)(Hb) = [\theta(Na)]\,[\theta(Nb)].$$

Its kernel is $\{Ng : Hg = H\} = \{Ng : g \in H\} = H/N$. Hence, by Theorem 1.20,

$$(G/N)/(H/N) \simeq G/H \,.$$

$\square$

Next, we have a straightforward result concerning abelian groups:

Theorem 9.8

Let A be a finite abelian group and let p be a prime such that p divides $|A|$. Then A contains an element of order p.

Proof

We use induction on $|A|$, noting that the result is trivial if $|A| = p$. Let $|A| = p^k n$, where $k \geq 1$ and $p \nmid n$. Let M be a maximal proper subgroup of A, with order m. If $p \mid m$ then, by induction, M (and hence, of course, A) contains an element of order p. So suppose that $p \nmid m$. Let $v \in A \setminus M$, and suppose that the cyclic subgroup $V = \langle v \rangle$ is of order r. Since MV is a subgroup of A properly containing M, we must have $MV = A$. From Theorem 9.7 we have that

$$A/M = MV/M \simeq V/(M \cap V)\,,$$

and so it follows that

$$p^k n = |A| = \frac{|M|\,|V|}{|M \cap V|} = \frac{mr}{|M \cap V|}\,.$$

Hence $p \mid r$, and so the element $v^{r/p}$ has order p. $\square$

As we shall see, this result holds also for non-abelian groups. The most convenient way to prove the more general result is to use a theorem due to Sylow, but before stating and proving the theorem we need to develop a little more theory.

Let G be a finite group, and let $a, b \in G$. We say that a is **conjugate** to b if there exists x in G such that $x^{-1}ax = b$. It is routine to check that conjugacy is an equivalence relation (see Exercise 9.5); hence G is partitioned into k equivalence classes C_i $(i = 1, 2, \ldots k)$. Within each C_i every element is conjugate to every other. It is clear that the only element conjugate to the identity element e is e itself, and we may suppose that $C_1 = \{e\}$. The **class equation** of G is the arithmetical equality deriving from the partition:

$$|G| = 1 + |C_2| + \cdots + |C_k|\,. \tag{9.9}$$

Remark 9.9

In an abelian group the notion of conjugacy is not useful, since elements are conjugate only if they are equal.

Let $a \in G$, and let $C(a)$ be the conjugacy class consisting of all elements that are conjugate to a. The **centraliser** $Z(a)$ is defined to be the set of all g in G such that $ga = ag$. It is easy (see Exercise 9.6) to verify that $Z(a)$ is a subgroup of G. There is a close connection between $C(a)$ and $Z(a)$:

Lemma 9.10

The number of elements in $C(a)$ is equal to the index of $Z(a)$ in G.

Proof

By the definition, $C(a) = \{x^{-1}ax : x \in G\}$. The elements $x^{-1}ax$ are not all distinct: $x^{-1}ax = y^{-1}ay$ if and only if $axy^{-1} = xy^{-1}a$, that is, if and only if $xy^{-1} \in Z(a)$, that is, if and only if x and y are in the same left coset of $Z(a)$. Thus the number of distinct elements in $C(a)$ is equal to the number of distinct cosets of $Z(a)$. $\qquad\square$

It is a consequence of this lemma that in the class equation each $|C_i|$ divides $|G|$.

The **centre** $Z = Z(G)$ of a group G is the set

$$\{z \in G : (\forall g \in G)\, zg = gz\}.$$

Alternatively, we can define Z as the set of elements z of G for which $Z(z) = G$. It is easy to verify that Z is a normal subgroup of G. Indeed (see Exercise 9.7) every subgroup U of G contained in $Z(G)$ is normal. Also immediate (see Exercise 9.7) is the result that $a \in Z$ if and only if $C(a) = \{a\}$.

The next result plays an important part in finite group theory:

Theorem 9.11

If G is a group of order p^m, where p is prime and m is a positive integer, then $Z(G)$ is non-trivial.

Proof

The class equation (9.9) gives

$$p^m = 1 + |C_2| + \cdots + |C_k|,$$

and so $1 + |C_2| + \cdots + |C_k|$ is divisible by p. Since each $|C_i|$ divides p^m, this can happen only if $|C_i| = 1$ for at least $p - 1$ values of i in $\{2, \ldots, k\}$. Hence $|Z(G| \geq p$. □

We are ready now to prove the result we need:

Theorem 9.12

Let G be a finite group of order $p^l r$, where p is prime and $p \nmid r$. Then G has at least one subgroup of order p^l.

Proof

We use induction on $|G|$, the result being clear if $|G| = 1$ or 2. Consider the class equation

$$p^l r = |G| = c_1 + c_2 + \cdots + c_k,$$

where $c_i = |C_i|$ $(i = 1, 2, \ldots, k)$. We know that that c_i is equal to $|G|/|Z_i|$, where Z_i is the centraliser in G of a typical element of C_i. If we write z_i for the order of Z_i, we obtain

$$z_i = \frac{p^l r}{c_i} \quad (i = 1, 2, \ldots, k). \tag{9.10}$$

Suppose first that there exists $c_i > 1$ such that $p \nmid c_i$. Then $z_i < p^l r$ and is divisible by p^l. Hence, by induction, Z_i contains a subgroup of order p^l, and we are home. We may therefore suppose that, for all i in $\{1, 2, \ldots, k\}$, either $c_i = 1$ or p divides c_i. The union of the classes C_i such that $c_i = 1$ is the centre Z of the group G (see Exercise 9.7) and so

$$p^l r = |Z| + vp$$

for some integer v. Hence Z is non-trivial, with order divisible by p. But Z is abelian and so, by Theorem 9.8, it contains an element a of order p. Since Z is normal, the cyclic subgroup $\langle a \rangle$ is certainly normal, and $|G/\langle a \rangle| = p^{l-1} r$. By the induction hypothesis, $G/\langle a \rangle$ contains a subgroup $U/\langle a \rangle$ of order p^{l-1}, and so G contains a subgroup U of order p^l. The subgroup U is called a **Sylow subgroup**. □

The following corollary, an easy consequence of Sylow's theorem, was proved earlier by Cauchy:

Corollary 9.13

Let G be a finite group and let p be a prime such that p divides $|G|$. Then G contains an element of order p.

Proof

We have seen that G has a subgroup H of order p^l. A typical element v of H has order p^k, where $k \leq l$, and it is then clear that $v^{p^{k-1}}$ has order p.   □

Remark 9.14

Theorem 9.12 is only part of Sylow's theorem – the only part we shall require. For the full result, see [13].

In Chapter 11, when we come to consider further applications to geometry, we shall need the following result:

Theorem 9.15

Let G be a group of order p^m, where p is prime and m is a positive integer. Then there exist normal subgroups

$$\{e\} = H_0 \subset H_1 \subset \cdots \subset H_{m-1} \subset H_m = G$$

of G such that $|H_i| = p^i$ for $i = 0, 1, \ldots, m$.

Proof

First, observe that G must contain an element of order p; for the order of an arbitrarily chosen $a \neq e$ in G is p^r for some r in $\{1, 2, \ldots, m\}$, and so $a^{p^{r-1}}$ is of order p.

For $m = 1$ there is nothing to prove. So let $m \geq 2$, suppose inductively that the result holds for all $k < m$, and let $|G| = p^m$. By Theorem 9.11, we may suppose that there is a subgroup P of order p contained in the centre $Z(G)$. We know that P is normal (see Exercise 9.7) and we have arranged that $|G/P| = p^{m-1}$. Every normal subgroup $\overline{N}$ of G/P may be written (see Exercise 1.31) as N/P, where N is a normal subgroup of G containing P. By the induction hypothesis, there exist normal subgroups K_i, all containing P, such that

$$\{e\} = K_0/P \subset K_1/P \subset \cdots \subset K_{m-1}/P = G/P,$$

with $|K_i/P| = p^i$ $(i = 1, 2, \ldots, m - 1)$. If we define $H_0 = \{e\}$, $H_1 = P$ and $H_i = K_{i-1}$ $(i = 2, \ldots, m)$, we obtain normal subgroups H_i of G such that

$$\{e\} = H_0 \subset H_1 \subset \cdots \subset H_{m-1} \subset H_m = G,$$

with $|H_i| = p^i$ $(i = 0, 1, \ldots, m)$. $\qquad\qquad\qquad\qquad\qquad\qquad\qquad\square$

EXERCISES

9.5. Show that conjugacy in a group is an equivalence relation.

9.6. Let a be an element of a group G. Show that the centraliser $Z(a)$ is a subgroup.

9.7. Let G be a group with centre Z.

   (i) Show that Z is a subgroup of G.

   (ii) Let H be a subgroup of G such that $H \subseteq Z$. Show that H is normal.

   (iii) Show that $a \in Z$ if and only if $C(a) = \{a\}$.

9.3 Permutation Groups

Let S_n be the **symmetric group on n symbols**, consisting of all one–one mappings (permutations) of the set $\{1, 2, \ldots, n\}$ onto itself, the operation being composition of mappings. It is useful (and traditional) to refer to the composition of two permutations π_1 and π_2 as their **product** and to interpret $\pi_1\pi_2$ as "first π_1, then π_2". This is equivalent to writing mapping symbols on the right. A **cycle** of length k, written $\sigma = (a_1 \; a_2 \; \ldots \; a_k)$ is a permutation such that

$$a_1\sigma = a_2, \; a_2\sigma = a_3, \ldots, \; a_{k-1}\sigma = a_k, \; a_k\sigma = a_1$$

and $x\sigma = x$ for each x not in the set $\{a_1, a_2, \ldots, a_k\}$.

Theorem 9.16

Every π in S_n can be expressed as a product of disjoint cycles. The order of π is the least common multiple of the lengths of the cycles.

Proof

Let x_1 be an arbitrarily chosen element of $\{1, 2, \ldots, n\}$. If $x_1\pi = x_1$, then (x_1) is itself a cycle; otherwise write $x_1\pi$ as x_2. We continue with a sequence

$$x_1, \; x_2 = x_1\pi, \; x_3 = x_2\pi, \ldots,$$

and, since the set $\{1, 2, \ldots, n\}$ is finite, there must eventually be a repetition: suppose that the first repetition is $x_k\pi = x_j$, with $k > j$. If $j \neq 1$ we have a contradiction, since $x_{j-1}\pi = x_k\pi = x_j$; hence $j = 1$, and the restriction of π to $\{x_1, x_2, \ldots, x_k\}$ is the cycle $(x_1 \; x_2 \; \ldots \; x_k)$.

Then choose y_1 not in $\{x_1, x_2, \ldots, x_k\}$ and repeat the process, obtaining a cycle $(y_1 \; y_2 \ldots y_l)$. Eventually the process must cease, and we obtain the decomposition of π into disjoint cycles.

It is clear that the order of a cycle coincides with its length, and that disjoint cycles commute with each other. Hence, if π is the product $\sigma_1\sigma_2\ldots\sigma_r$ of disjoint cycles of lengths $\lambda_1, \lambda_2, \ldots, \lambda_r$, then, for each $m \geq 1$,

$$\pi^m = \sigma_1^m\sigma_2^m\ldots\sigma_r^m\,,$$

and this is equal to the identity permutation if and only if m is a multiple of each of the integers $\lambda_1, \lambda_2, \ldots, \lambda_r$. $\qquad\qquad\square$

Remark 9.17

The decomposition into disjoint cycles is in effect unique. The cycles can begin with any one of their entries, and the order of the cycles is arbitrary, but this is the limit of the variability: for example, we may rewrite $(1 \; 4 \; 5)(2 \; 3)$ as $(3 \; 2)(4 \; 5 \; 1)$, but the basic structure cannot be changed.

A cycle of length 2 is called a **transposition** The following important result is easily deduced from Theorem 9.16:

Corollary 9.18

Every permutation can be expressed as a product of transpositions.

Proof

In view of Theorem 9.16, we need only show that a cycle is a product of transpositions. It is easy to verify that

$$(a_1 \; a_2 \; \ldots \; a_k) = (a_1 \; a_2)(a_1 \; a_3)\ldots(a_1 \; a_k)\,.$$

$\qquad\qquad\square$

The next observation is that permutations come in two different kinds: **even** and **odd**. There are various ways of defining these terms: perhaps the best is to consider the polynomial

$$\Delta(X_1, X_2, \ldots, X_n) = \prod_{1 \le i < j \le n} (X_i - X_j)$$

$$= (X_1 - X_2)\,(X_1 - X_3)\,\ldots\,(X_1 - X_n)$$
$$(X_2 - X_3)\,\ldots\,(X_2 - X_n)$$
$$\ldots\ldots$$
$$(X_{n-1} - X_n)\,.$$

in n indeterminates. (The polynomial has degree $(n-1) + (n-2) + \cdots + 1 = \frac{1}{2}n(n-1)$.) For each permutation π in the symmetric group S_n, we may define

$$\pi(\Delta) = \prod_{1 \le i < j \le n} (X_{\pi(i)} - X_{\pi(j)})\,.$$

The factors in $\pi(\Delta)$ are the same as the factors in Δ, except that they are in a different order, and some of them may be reversed. For example, if π is the transposition $(1\ 2)$, the factor $X_1 - X_2$ becomes $X_2 - X_1$, and all other factors are the same as before; thus $\pi(\Delta) = -\Delta$. On the other hand, if

$$\pi = (1\ 2\ 3) = (1\ 2)(1\ 3)\,,$$

two factors, namely, $X_1 - X_2$ and $X_1 - X_3$, change signs, and so $\pi(\Delta) = \Delta$. A permutation π is **even** or **odd** according as $\pi(\Delta) = \Delta$ or $\pi(\Delta) = -\Delta$. We see also that π is even [odd] if and only if it is expressible as a composition of an even [odd] number of transpositions. From this it follows that

$$\text{even} \circ \text{even} = \text{even}\,, \quad \text{even} \circ \text{odd} = \text{odd} \circ \text{even} = \text{odd}\,, \quad \text{odd} \circ \text{odd} = \text{even}\,.$$

Consequently the set of all even permutations is a subgroup, indeed a normal subgroup, of S_n, called the **alternating group**, and denoted by A_n.

The coset $A_n(x_1\ x_2)$ consists entirely of odd permutations. Moreover, every odd permutation π can be written as $\big(\pi(x_1\ x_2)\big)(x_1\ x_2)$, and $\pi(x_1\ x_2)$ is even. Thus the set of odd permutations is precisely $A_n(x_1\ x_2)$, and we conclude that A_n is of index 2 in S_n, and of order $\frac{1}{2}n!$.

Theorem 9.19

The symmetric group S_3 is soluble.

Proof

S_3 consists of the permutations

$$e = 1, \ a = (1 \ 2 \ 3), \ b = (1 \ 3 \ 2),$$

$$x = (2 \ 3), \ y = (1 \ 3), \ z = (1 \ 2).$$

S_3 has a normal subgroup $H = \{e, a, b\}$, and both H and S/H are cyclic. Thus S_3 is soluble.   □

The next result is not so immediate:

Theorem 9.20

The symmetric group S_4 is soluble.

Proof

The alternating group A_4 is a subgroup of index 2 and is certainly normal (see Exercise 1.26). Moreover, the quotient S_4/A_4, being a group of order 2, is assuredly abelian. The alternating group consists of the identity, together with the 11 elements

$$(1 \ 2 \ 3), \ (1 \ 2 \ 4), \ (1 \ 3 \ 2), \ (1 \ 3 \ 4), \ (1 \ 4 \ 2), \ (1 \ 4 \ 3), \ (2 \ 3 \ 4), (2 \ 4 \ 3),$$

$$(1 \ 2)(3 \ 4), \ (1 \ 3)(4 \ 4), \ (1 \ 4)(2 \ 3).$$

The set $V = \{1, (1 \ 2)(3 \ 4), (1 \ 3)(2 \ 4), (1 \ 4)(2 \ 3)\}$ is an abelian subgroup of A_4 (the Klein 4-group). Its right and left cosets are (after some computation)

$$V, \quad V(1 \ 2 \ 3) = (1 \ 2 \ 3)V = \{(1 \ 2 \ 3), (1 \ 3 \ 4), (1 \ 4 \ 2), (2 \ 4 \ 3)\},$$

$$V(1 \ 2 \ 4) = (1 \ 2 \ 4)V = \{(1 \ 2 \ 4), (1 \ 3 \ 2), (1 \ 4 \ 3), (2 \ 3 \ 4)\},$$

and so $V \lhd A_4$. The quotient A_4/V, being of order 3, is abelian. We thus have $1 \subseteq V \subseteq A_4 \subseteq S_4$, with $1 \lhd V$, $V \lhd A_4$, $A_4 \lhd S_4$; and $V/1$, A_4/V, S_4/A_4 are all abelian, and so, by Exercise 9.4, S_4 is soluble.   □

To determine whether S_n is soluble for $n \geq 5$, it is useful to look at the alternating group A_n. We begin with a lemma:

Lemma 9.21

For all $n \geq 3$, the alternating group A_n is generated by the set of all cycles of length 3.

Proof

It is clear that A_n is generated by the set of elements of type $(a\ b)(c\ d)$. If the two transpositions are equal, their product is the identity. If the product is of the form $(a\ b)(a\ c)$, where a, b, c are distinct, then we see that

$$(a\ b)(a\ c) = (a\ b\ c)\,;$$

and if a, b, c, d are all distinct, then

$$(a\ b)(c\ d) = [(a\ b)(a\ c)]\,[(c\ a)(c\ d)] = (a\ b\ c)(c\ a\ d)\,.$$

$\square$

A non-abelian group is called **simple** if it has no proper normal subgroups. Such a group is certainly not soluble. The systematic study of simple groups is, despite their name, far from easy, but fortunately we shall need only one result:

Theorem 9.22

For all $n \geq 5$, the alternating group A_n is simple.

Proof

Let $N \neq \{1\}$ be a normal subgroup of A_n; we shall show that N contains every cycle of length 3, and it will follow from Lemma 9.21 that $N = A_n$.

Case 1.   Suppose first that N contains a cycle $(a\ b\ c)$ of length 3. Let $x\ y,\ z$ be distinct elements in $\{1, 2, \ldots, n\}$, and let

$$\alpha = \begin{pmatrix} a & b & c \\ x & y & z \end{pmatrix}\,.$$

Then $\alpha^{-1}(a\ b\ c)\alpha = (x\ y\ z)$. If α is even, this implies that $(x\ y\ z) \in N$; if α is odd, we replace it by the even permutation $\beta = (d\ e)\alpha$, where $d, e \notin \{a, b, c\}$ (this being possible since $n \geq 5$) and observe that $\beta^{-1}(a\ b\ c)\beta = (x\ y\ z)$. Hence N contains all cycles of length 3, and so $N = A_n$.

Case 2.   Next, suppose that N contains an element π which decomposes into disjoint cycles as

$$\pi = \kappa_1\kappa_2\ldots\kappa_r\,,$$

and suppose that one of the cycles, which we may, without loss of generality, take as κ_1, is of length $s \geq 4$:

$$\kappa_1 = (a_1\ a_2\ \ldots\ a_s)\,.$$

Let $\alpha = (a_1 \ a_2 \ a_3)$; then $\alpha^{-1}\pi\alpha = (\alpha^{-1}\kappa_1\alpha)\kappa_2 \ldots \kappa_r$, since only κ_1 is affected by the conjugation, and

$$\alpha^{-1}\kappa_1\alpha = (a_1 \ a_3 \ a_2)(a_1 \ a_2 \ \ldots \ a_s)(a_1 \ a_2 \ a_3)$$
$$= (a_2 \ a_3 \ a_1 \ a_4 \ a_5 \ \ldots \ a_s).$$

The element $\pi^{-1}\alpha^{-1}\pi\alpha$ belongs to N, and

$$\pi^{-1}\alpha^{-1}\pi\alpha = \kappa_1^{-1}\alpha^{-1}\kappa_1\alpha$$
$$= (a_s \ a_{s-1} \ \ldots \ a_1)(a_2 \ a_3 \ a_1 \ a_4 \ a_5 \ \ldots \ a_s)$$
$$= (a_1 \ a_2 \ a_4).$$

We are back in Case 1, and so $N = A_n$.

It remains to consider the case where all the elements of N have cycle decompositions involving only cycles of length 2 and 3. If π contains only one cycle $(a \ b \ c)$ of length 3 (the other cycles being of length 2), then $\pi^2 = (a \ c \ b) \in N$, and we are back in Case 1. So suppose that π contains at least two disjoint cycles $(a \ b \ c)$ and $(d \ e \ f)$ of length 3. Then N contains

$$\pi' = (e \ d \ c)\pi(e \ c \ d)$$
$$= (e \ d \ c)(a \ b \ c)(d \ e \ f)(e \ c \ d)\ldots$$
$$= (a \ b \ d)(c \ f \ e)\ldots ,$$

and so contains

$$\pi\pi' = (a \ b \ c)(d \ e \ f)\ldots(a \ b \ d)(c \ f \ e)\ldots$$
$$= (a \ d \ c \ b \ f)\ldots .$$

We are back in Case 2, and so $N = A_n$.

The final case is where π is a product of a (necessarily even) number of transpositions. Suppose first that there are just two: $\pi = (a \ b)(c \ d)$. Then there is at least one other symbol e, since we are assuming that $n \geq 5$, and N contains the element

$$\pi[(a \ b \ e)^{-1}\pi(a \ b \ e)] = (a \ b)(c \ d)(a \ e \ b)(a \ b)(c \ d)(a \ b \ e)$$
$$= (a \ e \ b).$$

Again we are back in Case 1.

Suppose finally that $\pi = (a \ b)(c \ d)(e \ f)(g \ h)\ldots$. Then N contains

$$\pi[(b \ c)^{-1}(d \ e)^{-1}\pi(d \ e)(b \ c)] = \pi(b \ c)(d \ e)\pi(d \ e)(b \ c)$$
$$= (a \ e \ d)(b \ c \ f)\ldots ,$$

and once again we are back in a case already considered. $\square$

On the topic of symmetric groups, we shall require the following result:

Theorem 9.23

The symmetric group S_n is generated by the two cycles $(1\ 2)$ and $(1\ 2\ldots n)$.

Proof

Denote $(1\ 2)$ by τ and $(1\ 2\ldots n)$ by ζ. Then

$$\zeta^{-1} = \zeta^{n-1} = (n\ \ n-1\ldots 1),$$

and so

$$\zeta^{-1}\tau\zeta = (n\ \ n-1\ldots 1)(1\ 2)(1\ 2\ldots n) = (2\ 3),$$

and, more generally, we can show that

$$\zeta^{-i+1}\tau\zeta^{i-1} = (i\ \ i+1).$$

To see this, note that if $j \notin \{i, i+1\}$, then $(\bmod\ n)$

$$j\,\zeta^{-i+1}\tau\zeta^{i-1} = (j-i+1)\tau\zeta^{i-1} = (j-i+1)\zeta^{i-1} = j,$$

while

$$i\,\zeta^{-i+1}\tau\zeta^{i-1} = 1\tau\zeta^{i-1} = 2\zeta^{i-1} = i+1,$$

and

$$(i+1)\,\zeta^{-i+1}\tau\zeta^{i-1} = 2\tau\zeta^{i-1} = 1\zeta^{i-1} = i.$$

Next, observe that, for $j = 2, 3, \ldots, n-1$,

$$(j\ \ j+1)(j-1\ \ j)\ldots(2\ 3)(1\ 2)(2\ 3)\ldots(j\ \ j+1) = (1\ \ j+1).$$

We next show that, for $i = 1, 2, \ldots, n-1$ and $j = 1, 2, \ldots n-i$,

$$\zeta^{-i+1}(1\ \ j+1)\zeta^{i-1} = (i\ \ i+j);$$

for

$$i\,\zeta^{-i+1}(1\ \ j+1)\zeta^{i-1} = 1(1\ \ j+1)\zeta^{i-1} = (j+1)\zeta^{i-1} = i+j,$$

$$(i+j)\,\zeta^{-i+1}(1\ \ j+1)\zeta^{i-1} = (j+1)(1\ \ j+1)\zeta^{i-1} = 1\zeta^{i-1} = i,$$

and all other members of $\{1, 2, \ldots, n\}$ map to themselves.

We have shown that τ and ζ generate all transpositions in S_n, and it follows from Corollary 9.18 that they generate the whole of S_n. $\qquad\square$

9.4 Properties of Soluble Groups

Recall that a group G is **soluble** if, for some $m \geq 0$, it has a finite series

$$\{1\} = G_0 \subseteq G_1 \subseteq \cdots \subseteq G_m = G \qquad (9.11)$$

of subgroups such that, for $i = 0, 1, \ldots m - 1$,

(i) $G_i \triangleleft G_{i+1}$

(ii) G_{i+1}/G_i is cyclic.

Theorem 9.24

Let G be a group.

(i) If G is soluble, then every subgroup of G is soluble.

(ii) If G is soluble and N is a normal subgroup of G, then G/N is soluble.

(iii) Let $N \triangleleft G$. Then G is soluble if and only if both N and G/N are soluble.

Proof

(i) Suppose that

$$1 = G_0 \triangleleft G_1 \triangleleft \cdots \triangleleft G_m = G, \qquad (9.12)$$

and that G_{i+1}/G_i is cyclic for $i = 1, 2, \ldots, m - 1$. Let H be a subgroup of G, and, for each i, let $K_i = H \cap G_i$. Then, since $G_i \subseteq G_{i+1}$,

$$K_i = H \cap (G_{i+1} \cap G_i) = (H \cap G_{i+1}) \cap G_i = K_{i+1} \cap G_i.$$

From Theorem 9.7(i) it follows that $K_i \triangleleft K_{i+1}$, and that

$$K_{i+1}/K_i = K_{i+1}/(K_{i+1} \cap G_i) \simeq K_{i+1}G_i/G_i.$$

Since $K_{i+1}G_i/G_i$ is a subgroup of the cyclic group G_{i+1}/G_i, it is cyclic (or trivial), and so the sequence

$$\{1\} = K_0 \triangleleft K_1 \triangleleft \cdots \triangleleft K_m = H$$

has the required properties.

(ii) With G defined as before, it is clear that G/N has a series

$$N/N = G_0N/N \triangleleft G_1N/N \triangleleft \cdots \triangleleft G_mN/N = G/N.$$

(There may be coincidences in this series – for example, if $G_1 \subseteq N$, then $G_1 N / N = N/N$ – but this causes no problem.) Using Theorem 9.7 we can transform a typical quotient:

$$\frac{G_{i+1}N/N}{G_i N/N} \simeq \frac{G_{i+1}N}{G_i N} = \frac{G_{i+1}(G_i N)}{G_i N} \simeq \frac{G_{i+1}}{G_{i+1} \cap (G_i N)} \simeq \frac{G_{i+1}/G_i}{\big(G_{i+1} \cap (G_i N)\big)/G_i} \, .$$

The quotient, being isomorphic to a factor group of the cyclic group G_{i+1}/G_i is certainly cyclic.

(iii) From Parts (i) and (ii), we know that N and G/N are soluble if G is soluble.

Suppose conversely that N and G/N are soluble. Then there is a series

$$\{1\} = N_0 \lhd N_1 \lhd \cdots \lhd N_p = N \, ,$$

in which N_{i+1}/N_i is cyclic for $i = 0, 1, \ldots, p-1$, and a series

$$\{1\} = N/N = G_0/N \lhd G_1/N \lhd \cdots \lhd G_m/N = G/N \, ,$$

such that, for $i = 0, 1, \ldots, m-1$, $G_i \lhd G_{i+1}$ and $G_{i+1}/G_i \simeq (G_{i+1}/N)/(G_i/N)$ is cyclic. Hence there is a series

$$\{1\} = N_0 \lhd N_1 \lhd \cdots \lhd N_p = N = G_0 \lhd G_1 \lhd \cdots \lhd G_p = G \, ,$$

and so G is soluble. $\qquad\qquad\qquad\qquad\qquad\qquad\qquad\qquad\qquad\qquad\qquad$ $\square$

Corollary 9.25

For all $n \geq 5$, the symmetric group S_n is not soluble.

Proof

If S_n were soluble, then all its subgroups would be soluble, and we know that A_n, being simple, is certainly not soluble. $\qquad\qquad\qquad\qquad\qquad\qquad$ $\square$

10
Groups and Equations

The use of the word "soluble" for the groups described in the last chapter is on the face of it rather strange, but you may have guessed that there is a connection with equations and solubility by radicals. The following theorem establishes half of the connection:

Theorem 10.1

Let K be a field of characteristic zero. Let f be a polynomial in $K[X]$ whose Galois group $\mathrm{Gal}(f)$ is soluble. Then f is soluble by radicals.

Proof

Let L be a splitting field of f over K. We are supposing that $\mathrm{Gal}(L:K)$ is soluble; suppose also that $|\mathrm{Gal}(L:K)| = m$. If K does not contain an mth root of unity, we can certainly adjoin one: let E be the splitting field over K of the polynomial $X^m - 1$. Now let M be a splitting field for f over E.

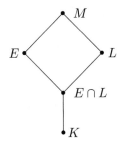

By Theorem 7.36, we may regard M as an extension of L, and $\mathrm{Gal}(M : E) \simeq \mathrm{Gal}(L : E \cap L)$. Now $\mathrm{Gal}(L : E \cap L)$ is a subgroup of the soluble group $\mathrm{Gal}(L : K)$, and so, by Theorem 9.24, $G = \mathrm{Gal}(M : E)$ is soluble. That is, there exist subgroups

$$\{1\} = G_0 \triangleleft G_1 \triangleleft \cdots \triangleleft G_r = G$$

such that G_{i+1}/G_i is cyclic for $0 \le i \le r - 1$. By the fundamental theorem, there is a corresponding sequence

$$E = M_r \subseteq M_{r-1} \subseteq \cdots \subseteq M_0 = M$$

of subfields of M, such that $\mathrm{Gal}(M : M_i) = G_i$, and $\mathrm{Gal}(M_i : M_{i+1}) \simeq G_{i+1}/G_i$. Thus M_i is a cyclic extension of M_{i+1}.

Let $[M_i : M_{i+1}] = d_i$ $(i = 0, 1, \ldots, r)$. Then d_i divides $[M : E] = |\mathrm{Gal}(M : E)|$, which in turn divides $|\mathrm{Gal}(L : K)| = m$. Since M_{i+1} contains E, it contains every mth root ω of unity, and so certainly contains all d_ith roots of unity, these being powers of ω. Hence, by Theorem 8.20 there exists an element β_i in M_i such that $M_i = M_{i+1}(\beta_i)$, where β_i is a root of an irreducible polynomial $X^{d_i} - c_{i+1}$, with c_{i+1} in M_{i+1}. It follows that the polynomial f is soluble by radicals. $\qquad\square$

The converse result is as follows:

Theorem 10.2

Let K be a field of characteristic zero, and let $K \subseteq L \subseteq M$, where M is a radical extension. Then $\mathrm{Gal}(L : K)$ is a soluble group.

Proof

We are supposing that there is a sequence

$$K = M_0, M_1, \ldots, M_r = M$$

such that $M_{i+1} = M_i(\alpha_i)$ $(i = 0, 1, \ldots, r-1)$, where α_i is a root of a polynomial $X^{n_i} - a_i$, irreducible in $M_i[X]$.

The *idea* of the proof is simple enough. At each stage the element α_i is a root of $X^{n_i} - b_i$, we use Theorems 8.13 and 8.18 to get useful information about the Galois groups. Unfortunately, we have to be careful that we have normal extensions at each stage. Right at the beginning of the argument, we see that L need not be a normal extension of K, and we need some preliminary repair work. We do this not by repairing L, but by modifying the base field K.

In the notation of Section 7.2, the fixed field $\Phi\big(\Gamma(K)\big)$ of $\mathrm{Gal}(L:K)$ will in general be larger than K: let us denote it by K'. (See Theorem 7.11.) On the other hand (see Exercise 7.3) we know that

$$\Phi\big(\Gamma(K')\big) = (\Phi\Gamma\Phi\Gamma)(K) = (\Phi\Gamma)(K) = K',$$

and so, by Theorem 7.30, L is a normal extension of K'. Since any polynomial f in $K[X]$ may certainly be regarded as a polynomial in $K'[X]$, and since $\mathrm{Gal}(L:K) = \mathrm{Gal}(L:K')$, we may replace K by K'. To avoid proliferation of notation, let us in fact suppose that L is a normal extension of K.

If N is a normal closure of M, then N is a radical extension, by Theorem 8.4. So there is no loss of generality in the statement of the theorem if we make the stronger assumption that M is both radical and normal.

If we prove that $\mathrm{Gal}(M:K)$ is soluble, it will follow from Theorems 7.34 and 9.24 that $\mathrm{Gal}(L:K)$ is soluble, for $\mathrm{Gal}(M:L) \lhd \mathrm{Gal}(M:K)$, and

$$\mathrm{Gal}(L:K) \simeq \mathrm{Gal}(M,K)/\mathrm{Gal}(M,L) \,.$$

So we set out to prove that $\mathrm{Gal}(M:K)$ is soluble, our assumption being that M is a normal (separable) radical extension of K.

Let

$$M = K(\alpha_1, \alpha_2, \ldots, \alpha_n) \,, \tag{10.1}$$

and suppose that

$$\alpha_i^{p_i} \in K(\alpha_1, \alpha_2, \ldots, \alpha_{i-1}), \quad (i = 1, 2, \ldots, n) \,.$$

We may in fact suppose that p_i is prime for all i, at a cost of increasing n: if, for example, we have $\alpha_i^{pq} \in K(\alpha_1, \alpha_2, \ldots, \alpha_n)$, we can define β as α_i^p, and say that $\beta^q \in K(\alpha_1, \alpha_2, \ldots, \alpha_n)$, and $\alpha_i^p \in K(\beta, \alpha_1, \alpha_2, \ldots, \alpha_n)$.

We prove the result by induction on the integer n featuring above in (10.1). We have that $\alpha_1^{p_1} = b_1 \in K$. In order to have enough roots of unity, we let $P = M(\omega)$ be a splitting field for $X^{p_1} - 1$ over M, where ω is a primitive p_1th root of unity. Certainly, by Theorem 7.13, P is a normal extension of M, and so, by Theorem 7.34, $\mathrm{Gal}(P:M) \lhd \mathrm{Gal}(P:K)$, and

$$\mathrm{Gal}(M:K) \simeq \mathrm{Gal}(P:K)/\mathrm{Gal}(P:M) \,.$$

Theorem 9.24 now tells us that that if we can prove that $\mathrm{Gal}(P:K)$ is soluble, then the desired result, that $\mathrm{Gal}(M:K)$ is soluble, will follow immediately.

Let M_1 be the subfield $K(\omega)$ of P. In fact M_1 is a splitting field over K of $X^{p_1} - 1$, and so is a normal extension. By Corollary 8.14, $\mathrm{Gal}(M_1:K)$ is cyclic (and hence certainly soluble). Thus $\mathrm{Gal}(P:M_1) \lhd \mathrm{Gal}(P:K)$, and

$$\mathrm{Gal}(M_1:K) \simeq \mathrm{Gal}(P:K)/\mathrm{Gal}(P:M_1) \,.$$

Hence once again Theorem 9.24 tells us that, if we can prove that $\mathrm{Gal}(P:M_1)$ is soluble, then solubility of $\mathrm{Gal}(P:K)$ will follow.

So, having begun with $\mathrm{Gal}(L:K)$, we have now reduced the problem to showing that $\mathrm{Gal}(P:M_1)$ is soluble. We may write

$$P = M_1(\alpha_1, \alpha_2, \ldots, \alpha_n)$$

(these being the same α_i as in (10.1)). Denote $\mathrm{Gal}(P:M_1)$ by G, and let $H = \mathrm{Gal}\big(P:M(\alpha_1)\big)$, a subgroup of G. The proof proceeds by induction on n.

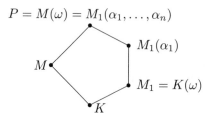

Now in $M_1[X]$,

$$X^{p_1} - 1 = (X-1)(X-\omega)(X-\omega^2)\ldots(X-\omega^{p_1-1}),$$

and in $\big(M(\alpha_1)\big)[X]$,

$$X^{p_1} - b_1 = X^{p_1} - \alpha_1^{p_1} = (X-\alpha_1)(X-\omega\alpha_1)(X-\omega^2\alpha_1)\ldots(X-\omega^{p_1-1}\alpha_1);$$

thus $M(\alpha_1)$ is a splitting field for $X^{p_1} - b_1$ over M_1. By Theorem 8.18, $\Gamma\big(M(\alpha_1)\big) = \mathrm{Gal}\big(M_1(\alpha_1):M_1\big)$ is cyclic. Since $M_1(\alpha)$ is a normal extension (being a splitting field) of M_1, it follows from Theorem 7.34 that $H \lhd G$ and that $G/H \simeq \Gamma\big(M(\alpha_1)\big)$ is cyclic.

Now H is the Galois group

$$\mathrm{Gal}\big(P:M(\alpha_1)\big) = \mathrm{Gal}\big(M_1(\alpha_1)(\alpha_2,\ldots,\alpha_n):M_1(\alpha_1)\big),$$

and P is a normal extension of $M_1(\alpha_1)$. The induction hypothesis allows us to assume that H is soluble. Since G/H is certainly soluble, we deduce that G is soluble. $\qquad\square$

Theorem 10.2 makes no reference to polynomials or equations, but this omission is easily repaired. Let f be a polynomial in $K[X]$, and suppose that it is soluble by radicals. That is to say, suppose that its splitting field L is contained in a radical extension M of K. The theorem tells us that $\mathrm{Gal}(f) = \mathrm{Gal}(L:K)$ is soluble. Theorems 10.2 and 10.1 together give the fundamental result:

Theorem 10.3

A polynomial f with coefficients in a field K of characteristic zero is soluble by radicals if and only if its Galois group is soluble.

10.1 Insoluble Quintics

To answer the classical question regarding quintic equations, we must now exhibit quintic polynomials in $\mathbb{Q}[X]$ with Galois groups that are not soluble. A rather technical observation gives us the means of constructing examples:

Theorem 10.4

Let p be a prime, and let f be a monic irreducible polynomial of degree p, with coefficients in $\mathbb{Q}$. Suppose that f has precisely two zeros in $\mathbb{C} \setminus \mathbb{R}$. Then the Galois group of f is the symmetric group S_p.

Proof

The polynomial f has a splitting field L contained in $\mathbb{C}$. The Galois group $G = \mathrm{Gal}(L : \mathbb{Q})$ is a group of permutations on the p roots of f in L, the roots being all distinct, by Theorem 7.22. Thus G is a subgroup of S_p. In constructing the splitting field of f, the first step is to form $\mathbb{Q}(\alpha)$, where α has minimum polynomial f. Then $[\mathbb{Q}(\alpha) : \mathbb{Q}] = p$. Since

$$p = \left| \mathrm{Gal}\big(\mathbb{Q}(\alpha) : \mathbb{Q}\big) \right| = \left| \mathrm{Gal}(L : \mathbb{Q}) \right| / \left| \mathrm{Gal}(L : \mathbb{Q}(\alpha)) \right|,$$

p divides $|G|$. By Theorem 9.12, G contains an element of order p. By Theorem 9.16, the only elements of order p in S_p are cycles of length p, and so G contains a cycle of length p.

The two non-real roots of f are necessarily complex conjugates of each other (see the proof of Theorem 2.20) and so the splitting field contains a transposition, interchanging the two non-real roots and leaving the rest unchanged. There is no loss of generality in denoting the transposition by $(1\ 2)$. We may also suppose that the p-cycle $\sigma = (a_1\ a_2 \ldots a_p)$ has $a_1 = 1$, for the choice of first element is arbitrary. If $a_k = 2$, then $\sigma^{k-1} = (1\ 2 \ldots)$, and we may as well write it as $(1\ 2 \ldots p)$. Since G contains $(1\ 2)$ and $(1\ 2 \ldots p)$, it follows from Theorem 9.23 that $G = S_p$. $\qquad\square$

Can we find an irreducible quintic polynomial with precisely 3 real roots? Yes, we can!

Example 10.5

Show that $f(X) = X^5 - 8X + 2$ is not soluble by radicals.

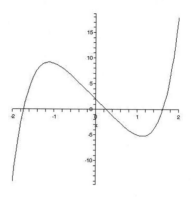

Figure 10.1. $X^5 - 8X + 2$

Solution

The polynomial f is certainly irreducible over $\mathbb{Q}$, by the Eisenstein criterion (Theorem 2.27). A table of values,

X	-2	-1	0	1	2
$f(X)$	-14	9	2	-5	18

tells us that there are roots in the intervals $(-2, -1)$, $(0, 1)$ and $(1, 2)$. The derivative $f'(X) = 5X^4 - 8$ is positive except between $-\sqrt[4]{8/5}$ and $\sqrt[4]{8/5}$, that is (approximately), between -1.1247 and 1.1247. By Rolle's theorem (see [7]) there is at least one zero of $f'(X)$ between zeros of $f(X)$, and so there are precisely three real roots. The graph of f is approximately as in Fig10.1. By Theorems 9.25 and 10.4, the polynomial $f(X)$ is not soluble by radicals.   □

EXERCISES

10.1. Show that $X^5 - 6X + 3$ is not soluble by radicals.

10.2. Show that $X^5 - 4X + 2$ is not soluble by radicals.

10.2 General Polynomials

Let K be a field of characteristic zero, and let L be an extension of K. A subset $\{\alpha_1, \alpha_2, \ldots, \alpha_n\}$ of L is said to be **algebraically independent** over K if, for all polynomials $f = f(X_1, X_2, \ldots, X_n)$ with coefficients in K,

$$f(\alpha_1, \alpha_2, \ldots, \alpha_n) = 0 \text{ only if } f = 0.$$

This is a much stronger condition than linear independence: the set

$$\{1, \sqrt{2}, \sqrt{3}, \sqrt{6}\}$$

is linearly independent over $\mathbb{Q}$, but is not algebraically independent, since (for example) the polynomial $f(X_1, X_2, X_3, X_4) = X_2 X_3 - X_4$ has

$$f(1, \sqrt{2}, \sqrt{3}, \sqrt{6}) = 0.$$

Indeed, algebraic independence can alternatively (see Exercise 8.16) be defined by the property that α_1 is transcendental over K and, for each r in $\{2, 3, \ldots, n\}$, α_r is transcendental over $K(\alpha_1, \alpha_2, \ldots, \alpha_{r-1})$. For yet another way of looking at the concept, $\{\alpha_1, \alpha_2, \ldots, \alpha_n\}$ is algebraically independent over K if and only if $K(\alpha_1, \alpha_2, \ldots, \alpha_n)$ is isomorphic to $K(X_1, X_2, \ldots, X_n)$, the field of all rational forms with n indeterminates and coefficients in K. (Again, see Exercise 8.16.)

An extension L of a field K is said to be **finitely generated** if, for some natural number m there exist elements $\alpha_1, \alpha_2, \ldots, \alpha_m$ such that $L = K(\alpha_1, \alpha_2, \ldots, \alpha_m)$. Every *finite* extension is certainly finitely generated, but the converse statement is false. The following theorem shows how to identify the transcendental aspect of a finitely generated extension:

Theorem 10.6

Let $L = K(\alpha_1, \alpha_2, \ldots, \alpha_n)$ be a finitely generated extension of K. Then there exists a field E such that $K \subseteq E \subseteq L$ such that, for some m such that $0 \leq m \leq n$:

(i) $E = K(\beta_1, \beta_2, \ldots, \beta_m)$, where $\{\beta_1, \beta_2, \ldots, \beta_m\}$ is algebraically independent over K;

(ii) $[L : E]$ is finite.

Proof

Suppose first that all of the elements $\alpha_1, \alpha_2, \ldots, \alpha_n$ are algebraic over K. Then $[L : K]$ is finite, and we may take $E = K$ and $m = 0$. Otherwise, there exists an α_i which is transcendental over K: let this be the element β_1. If $[L : K(\beta_1)]$ is not finite, there is an α_j which is transcendental over $K(\alpha_1)$; call it β_2. The process continues, and must terminate in at most n steps. Thus $E = K(\beta_1, \beta_2, \ldots, \beta_m)$, where $m \leq n$, $\{\beta_1, \beta_2, \ldots, \beta_m\}$ is algebraically independent over K, and $[L : E]$ is finite. $\qquad\square$

The elements β_i featuring in Theorem 10.6 are not unique, but m, the *number* of algebraically independent elements, is wholly determined by L and K:

Theorem 10.7

With the notation of Theorem 10.6, suppose that there is another field F such that $K \subseteq F \subseteq L$, and

(i) $F = K(\gamma_1, \gamma_2, \ldots, \gamma_p)$, where $\{\gamma_1, \gamma_2, \ldots, \gamma_p\}$ is algebraically independent over K;

(ii) $[L : F]$ is finite.

Then $p = m$.

Proof

Suppose that $p > m$. Since $[L : E]$ is finite, the element γ_1 is algebraic over E. It may even belong to E, but at worst γ_1 is a root of a polynomial with coefficients in $E = K(\beta_1, \beta_2, \ldots, \beta_m)$. Equivalently, there is a non-zero polynomial f such that

$$f(\beta_1, \beta_2, \ldots, \beta_m, \gamma_1) = 0.$$

Since the element γ_1 is transcendental over K, at least one of the elements β_i – say β_1 – must actually feature in the coefficients of the polynomial f. Thus β_1 is algebraic over $K(\beta_2, \ldots, \beta_m, \gamma_1)$, and $[L : K(\beta_2, \ldots, \beta_n, \gamma_1)]$ is finite. We can continue the argument, replacing each successive β_i by γ_i, and so $[L : K(\gamma_1, \gamma_2, \ldots, \gamma_m)]$ is finite. We are assuming that $p > m$, and we now have a contradiction, since γ_{m+1} is transcendental over $K(\gamma_1, \gamma_2, \ldots, \gamma_m)$. In exactly the same way, we obtain a contradiction if we assume that $m > p$, and the result follows. $\qquad\square$

The number m featuring in Theorem 10.6 is called the **transcendence degree** of L over K.

Let K be a field and let L be an extension of K with transcendence degree n. Let us suppose, in fact, that $L = K(t_1, t_2, \ldots, t_n)$, where $t_1, t_2, \ldots, t_n$ are algebraically independent over K. For all σ in the symmetric group S_n we can define a K-automorphism φ_σ of L, given by

$$\varphi_\sigma(t_i) = t_{\sigma(i)},$$

and extending in the usual way to L. Thus, for example, if $n = 3$ and $q = (t_1 + 3t_2 - t_3)/(t_1^3 t_2)$ and σ is the cycle $(1\ 2\ 3)$, then

$$\sigma(q) = (t_2 + 3t_3 - t_1)/(t_2^3 t_3).$$

Let us denote by Aut_n the group $\{\phi_\sigma : \sigma \in S_n\}$. The map $\sigma \mapsto \phi_\sigma$ is an isomorphism.

The fixed field F of Aut_n includes all the **elementary symmetric polynomials**

$$s_1 = t_1 + t_2 + \cdots + t_n\,,$$
$$s_2 = t_1 t_2 + t_1 t_3 + \cdots + t_{n-1} t_n\,,$$
$$\cdots$$
$$s_n = t_1 t_2 \ldots t_n\,;$$

and all rational combinations of these polynomials. For example, $t_1^2 + t_2^2 + \cdots + t_n^2$ is clearly in F, and a little elementary algebra establishes that it can be expressed as $s_1^2 - 2s_2$. The next theorem tells us that F is generated by $s_1, s_2, \ldots, s_n$:

Theorem 10.8

With the above notation, $F = K(s_1, s_2, \ldots, s_n)$.

Proof

We show, by induction on n, that

$$[K(t_1, t_2, \ldots, t_n) : K(s_1, s_2, \ldots, s_n)] \leq n!\,, \tag{10.2}$$

it being obvious that this holds for $n = 1$. Certainly

$$K(s_1, s_2, \ldots, s_n) \subseteq K(s_1, s_2, \ldots, s_n, t_n) \subseteq K(t_1, t_2, \ldots, t_n)\,.$$

The polynomial

$$f(X) = X^n - s_1 X^{n-1} + \cdots + (-1)^n s_n$$

factorises into $(X - t_1)(X - t_2) \ldots (X - t_n)$ over $K(t_1, t_2, \ldots, t_n)$. Hence the minimum polynomial of t_n over $K(s_1, s_2, \ldots, s_n)$ divides f. Consequently

$$[K(s_1, s_2, \ldots, s_n, t_n) : K(s_1, s_2, \ldots, s_n)] \leq n\,. \tag{10.3}$$

Let $s_1', s_2', \ldots, s_{n-1}'$ be the elementary symmetric polynomials in $t_1, t_2, \ldots, t_{n-1}$; then $s_1 = s_1' + t_n$, $s_n = s_{n-1}' t_n$, and

$$s_j = s_{j-1}' t_n + s_j' \quad (j = 2, 3, \ldots, n - 1)\,.$$

Hence

$$K(s_1, s_2, \ldots, s_n) = K(s_1', s_2', \ldots s_{n-1}', t_n)\,,$$

and so, by the induction hypothesis,

$$[K(t_1, t_2, \ldots, t_n) : K(s_1, s_2, \ldots, s_n, t_n)]$$
$$= [K(t_n)(t_1, t_2, \ldots, t_{n-1}) : K(t_n)(s'_1, s'_2, \ldots s'_{n-1})]$$
$$\leq (n-1)! \, .$$

This, together with (10.3), establishes (10.2).

Certainly it is clear that $K(s_1, s_2, \ldots, s_n)$ is contained in the fixed field F of Aut_n. By Theorem 7.12, $[K(t_1, t_2, \ldots, t_n) : F] = |\mathrm{Aut}_n| = n!$, and so, by (10.2), we must have $F = K(s_1, s_2, \ldots, s_n)$. $\square$

Theorem 10.9

The symmetric polynomials $s_1, s_2, \ldots, s_n$ are algebraically independent.

Proof

The field $F(t_1, t_2, \ldots, t_n)$ is a finite extension of $F(s_1, s_2, \ldots, s_n)$, since t_1, t_2, $\ldots$, t_n are the roots of

$$X^n - s_1 X^{n-1} + s_2 X^{n-2} - \cdots + (-1)^n s_n \, .$$

Thus $F(t_1, t_2, \ldots, t_n)$ and $F(s_1, s_2, \ldots, s_n)$ have the same transcendence degree, and so $s_1, s_2, \ldots, s_n$ are algebraically independent. $\square$

Let us now consider a set of n algebraically independent elements over a field K with characteristic zero. For reasons that will appear shortly, we shall name the elements as $s_1, s_2, \ldots, s_n$, but for the moment they are just arbitrarily chosen algebraically independent elements. The **general polynomial** of degree n "over K" (though its coefficients are in fact in $K(s_1, s_2, \ldots, s_n)$) is

$$X^n - s_1 X^{n-1} + s_2 X^{n-2} - \cdots + (-1)^n s_n \, . \tag{10.4}$$

We can call it a general (or generic) polynomial, because there is no algebraic connection among the coefficients.

Theorem 10.10

Let K be a field of characteristic zero, and let $g(X)$ be given by (10.4). Let M be a splitting field for g over $K(s_1, s_2, \ldots, s_n)$. Then the zeros $t_1, t_2, \ldots, t_n$ of g in M are algebraically independent over K, and the Galois group of M over $K(s_1, s_2, \ldots, s_n)$ is the symmetric group S_n.

Proof

The degree $[M : K(s_1, s_2, \ldots, s_n))]$ is finite, by Theorem 5.1, and so, over K, the transcendence degree of M is the same as that of $K(s_1, s_2, \ldots, s_n)$, namely, n. Since $M = K(t_1, t_2, \ldots, t_n)$, the elements $t_1, t_2, \ldots, t_n$ must be algebraically independent. From the identity

$$X^n - s_1 X^{n-1} + s_2 X^{n-2} - \cdots + (-1)^n s_n = (X - t_1)(X - t_2) \ldots (X - t_n)$$

we deduce that $s_1, s_2, \ldots, s_n$ are the elementary symmetric polynomials in $t_1, t_2, \ldots, t_n$. As we have seen above, Aut_n is a group of automorphisms of M, and its fixed field is $K(s_1, s_2, \ldots, s_n)$. By Theorem 7.12,

$$[M : K(s_1, s_2, \ldots, s_n)] = |\mathrm{Aut}_n| = |S_n| = n! \, .$$

Hence $\mathrm{Gal}\big(M : K(s_1, s_2, \ldots, s_n)\big) \simeq S_n$. $\qquad\square$

We immediately deduce the final result:

Theorem 10.11

If K is a field with characteristic zero and $n \geq 5$, the general polynomial (10.4) is not soluble by radicals.

EXERCISES

10.3. Let K be a field and let L be an extension of K containing a set $\{\alpha_1, \alpha_1, \ldots, \alpha_n\}$. Show that the following statements are equivalent:

(i) $\{\alpha_1, \alpha_2, \ldots, \alpha_n\}$ is algebraically independent over K;

(ii) α_1 is transcendental over K and, for each r in $\{2, 3, \ldots, n\}$, α_r is transcendental over $K(\alpha_1, \alpha_2, \ldots, \alpha_{r-1})$;

(iii) $K(\alpha_1, \alpha_2, \ldots, \alpha_n) \simeq K(X_1, X_2, \ldots, X_n)$.

10.4. Let α, a real number, be transcendental over $\mathbb{Q}$. Is it possible to find a real number β which is transcendental over $\mathbb{Q}(\alpha)$? [Hint: think of cardinal numbers.]

10.5. It follows from Theorem 10.8 that every symmetric polynomial is a rational expression in the elementary symmetric polynomials $s_1, s_2, \ldots, s_n$. Express $t_1^3 + t_2^3 + t_3^3$ as a rational expression in s_1, s_2, s_3.

10.3 Where Next?

Wisely, we confined ourselves in this chapter to fields with characteristic zero. The conclusions for fields of characteristic p are not hugely different, and are not a lot harder to prove. All that is necessary is to make sure that p does not get mixed up with the degrees of extensions and the order of Galois groups. We can state a theorem as follows:

Theorem 10.12

Let f be a separable polynomial in $K[X]$, where char $K = p$. If the Galois group $\mathrm{Gal}(f)$ is soluble, and if p does not divide $|\mathrm{Gal}(f)|$, then f is soluble by radicals.

Conversely, we have a more complicated statement:

Theorem 10.13

Let K be a field with prime characteristic p. Let $L : K$ be a Galois extension with subfields $K = L_0 \subset L_1 \subset \cdots \subset L_r = L$, and suppose that, for $i = 1, 2, \ldots, r$,

$$L_i = L_{i-1}(\alpha_i),$$

where α_i is a root of $X^{n_i} - c_i$, with $c_i \in L_{i-1}$. Suppose also that p does not divide $n_1 n_2 \ldots n_r$. If f splits over L, then the Galois group $\mathrm{Gal}(f)$ is soluble.

At the very beginning of this chapter we made the obvious comment that the roots of a polynomial $X^n + a_{n-1}X^{n-1} + \cdots + a_1 X + a_0$ are determined by the coefficients – are, to put it another way, functions $\rho(a_0, a_1, \ldots, a_{n-1})$. For $n = 1, 2, 3, 4$ the function ρ is what we might call "rational-radical", but we now know that this is not the case for $n \geq 5$. So what kind of function is it? Hermite showed that, for $n = 5$, the solution can be expressed in terms of **elliptic modular functions** (see [4]), functions that arise in quite a different context, and this work was developed by Klein[1] and Poincaré[2] .

In another direction, it was not long before an obvious question was asked. Given a finite group G, define G as **realisable** if there exists a polynomial in $\mathbb{Q}[X]$ having G as its Galois group. Which groups are realisable? A deep result, due to Shafarevich[3] in 1956, is that every soluble group is realisable.

---

[1] Felix Christian Klein, 1849–1925.
[2] Jules Henri Poincaré, 1854–1912.
[3] Igor Rostislavovich Shafarevich, 1923–.

(See [12].) At the other extreme, it is not known whether every finite simple group is realisable.

In the case of quintic polynomials, only 5 groups are realisable. We have come across two of them:

- the metacyclic group $M_{20} = \langle a, b \mid a^5 = b^4 = a^2 b a^{-1} b^{-1} = 1 \rangle$, which we encountered in Example 8.22;

- the symmetric group S_5, which we encountered in Example 10.5;

and the other three are

- the cyclic group C_5;

- the alternating group A_5; and

- the dihedral group $D_5 = \langle a, b \mid a^5 = b^2 = 1, \ ab = ba^4 \rangle$.

For $n > 5$ we have less information. In general, the calculation of Galois groups is quite difficult, and many questions remain unanswered. The topic is still very much alive: for example, as recently as 1987, Osada [11] showed that $\mathrm{Gal}(X^n - X - 1) \simeq S_n$ for all $n \geq 2$.

All of this, however, is well beyond the scope of an introductory text on Galois theory. For further information, see [5] and [9].

11
Regular Polygons

11.1 Preliminaries

After the undeniably hard work of the last two chapters, we reward ourselves with a "lollipop" by returning to the theme of constructions using ruler and compasses. The fact (to be established below) that a 17-sided regular polygon is constructible, whereas a 19-sided regular polygon is not, is of little practical significance, but the argument is beautiful, and, to the soul of a pure mathematician, is its own justification. Even the fact that a $65,537$-sided polygon is constructible is intriguing!

In Chapter 4 we used Theorem 4.8 to show the impossibility of certain constructions, the most celebrated being the problem of squaring the circle. In this chapter we wish also to demonstrate the *possibility* of certain constructions, and for this we need what amounts to a converse of Theorem 4.8.

It is convenient to begin with a lemma. Recall that a point (a, b) is **constructible** if it can be obtained from $O = (0, 0)$ and $I = (1, 0)$ by ruler and compasses constructions.

Lemma 11.1

Let $a, b \in \mathbb{R}$.

(i) The point $(a, 0)$ is constructible if and only if $(0, a)$ is constructible.

(ii) The point (a, b) is constructible if and only if $(a, 0)$ and $(b, 0)$ are constructible.

(iii) If $(a,0)$ and $(b,0)$ are constructible, then so are $(a+b,0)$, $(a-b,0)$, $(ab,0)$ and, if $b \neq 0$, $(a/b,0)$.

Proof

(i) Suppose that $(a,0)$ is constructible. The circle with centre O passing through $(a,0)$ meets the positive y-axis in $(0,a)$, and so $(0,a)$ is constructible. The converse is clear.

(ii) Suppose that (a,b) is constructible. Then, by Example 4.2 we can drop a perpendicular from (a,b) on to the x-axis to construct the point $(a,0)$. Dropping a perpendicular on to the y-axis gives the point $(0,b)$; and so, by Part (i), both $(a,0)$ and $(b,0)$ are constructible.

Conversely, suppose that $(a,0)$ and $(b,0)$ (and hence also $(0,b)$) are constructible. By Example 4.1 we may draw a line through $(a,0)$ perpendicular to the x-axis, and a line through $(0,b)$ perpendicular to the y-axis. The lines meet in (a,b), which is therefore constructible.

(iii) Suppose that $A = (a,0)$ and $B = (b,0)$ are constructible. A circle with centre A and radius equal to the length of OB meets the x-axis in $(a+b,0)$ and $(a-b,0)$. Hence both these points are constructible. To show that $(ab,0)$ is constructible, let $A' = (0,a)$ and $I' = (0,1)$ (both constructible, by Part (i)). By Example 4.5 we may draw a line though A' parallel to $I'B$, meeting the x-axis in P.

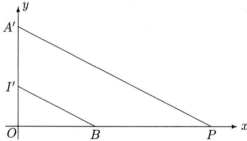

The triangles OBI' and OPA' are similar, and so $OP/OA' = OB/OI'$. Hence P is the point $(ab,0)$, and we have shown that it is constructible.

Finally, let B be the point $(b,0)$, where $b \neq 0$, and, as before, let $I' = (0,1)$.

Draw a line through I parallel to BI', meeting the y-axis in P.

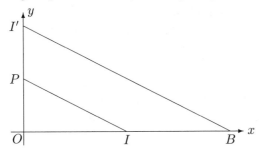

The triangles OIP and OBI' are similar, and so $OP/OI = OI'/OB$. Thus P is the point $(0, 1/b)$, and it follows that $(1/b, 0)$ is constructible. From the result of the last paragraph we may now deduce that $(a/b, 0)$ is constructible. □

Corollary 11.2

If $a, b \in \mathbb{Q}$, then (a, b) is constructible.

Proof

From Part (iii) of the lemma, we can deduce that $(m/n, 0)$ is constructible for every rational number m/n. Thus $(a, 0)$ and $(0, b)$ are constructible; and so, by Part (ii), (a, b) is constructible. □

We are ready now to prove the following converse to Theorem 4.8.

Theorem 11.3

Let $B = \{O, I\}$. If there is a sequence of subfields

$$\mathbb{Q} = K_0 \subset K_1 \subset \cdots \subset K_n = L$$

of $\mathbb{R}$ such that $[K_i : K_{i-1}] = 2$ $(i = 1, 2, \ldots, n)$, then every point with coordinates in L is constructible.

Proof

From Corollary 11.2, every (a, b) with coordinates in $\mathbb{Q} = K_0$ is constructible. Suppose inductively that $i \geq 1$ and that every point with coordinates in K_{i-1} is constructible. Since $[K_i : K_{i-1}] = 2$, we may conclude (see Exercise 3.5) that $K_i = K_{i-1}(\beta)$, where β is an arbitrarily chosen element of $K_i \setminus K_{i-1}$.

The minimum polynomial of β over K_{i-1} is of the form $X^2 + bX + c$, with $b, c \in K_{i-1}$ and with discriminant $\Delta = b^2 - 4c \geq 0$, since K_i is certainly a subfield of $\mathbb{R}$. Since $\beta = \frac{1}{2}(b \pm \sqrt{\Delta})$, it is clear that $K_i = K_{i-1}(\sqrt{\Delta})$, where $\Delta \in K_{i-1}$. All we need to do to complete the proof is to show that $(\sqrt{\Delta}, 0)$ is constructible.

Let D be the point $(\Delta, 0)$, and let E be the point on the x-axis such that $IE = \Delta$. Let M be the midpoint of OE, and let $\mathcal{K}$ be the circle with centre M passing through O (and E). Let the line through I perpendicular to the x-axis meet the circle $\mathcal{K}$ in P.

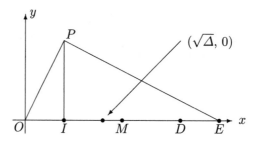

The angle OPE is a right angle, and the triangles OIP and PIE are similar; hence $OI/IP = IP/IE$. That is, $IP^2 = \Delta$. The point $(\sqrt{\Delta}, 0)$ is obtained as the intersection with the positive x-axis of a circle with centre O and radius equal to the length IP. It now follows from Lemma 11.1 that an arbitrary point $(p + q\sqrt{\Delta},\ r + s\sqrt{\Delta})$ where $p, q, r, s \in K_{i-1}$, is constructible. $\square$

The conditions on L in the statement of Theorem 11.3 imply that $[L : \mathbb{Q}]$ is a positive power of 2, and it is reasonable to ask whether this more compactly expressed condition is sufficient for constructibility. In fact Theorem 9.15 is exactly what we need:

Theorem 11.4

Let K be a normal extension of $\mathbb{Q}$ such that $[K : \mathbb{Q}] = 2^m$, where m is a positive integer. Then every point (α, β) in $K \times K$ is constructible.

Proof

The group $G = \mathrm{Gal}(K, \mathbb{Q})$ is of order 2^m and, by Theorem 9.15, there exist normal subgroups

$$\{e\} = H_0 \subset H_1 \subset \cdots \subset H_{m-1} \subset H_m = G$$

such that $|H_i| = 2^i$ $(i = 0, 1, \ldots, m)$. By Theorem 7.34, there exist subfields

$$K = \Phi(H_0) \supset \Phi(H_1) \supset \cdots \supset \Phi(H_{m-1}) \supset \Phi(H_m) = \mathbb{Q},$$

with $[K : \Phi(H_i)] = 2^i$ $(i = 0, 1, \ldots, m)$. Hence $[\Phi(H_i) : \Phi(H_{i+1})] = 2$ $(i = 0, 1, \ldots, m - 1)$, and the conclusion now follows from Theorem 11.3. $\qquad\square$

11.2 The Construction of Regular Polygons

For all $n \geq 3$, denote the regular polygon with n sides by Π_n. In Chapter 4 we saw how to construct Π_n for $n = 4$; and some other small values of n, known to Euclid and his contemporaries, present no great problem. Gauss, aged 19 at the time, "out-Greeked the Greeks" by showing that Π_{17} is constructible, and it is said that his delight in this result convinced him that his future lay in mathematics. The techniques we have developed enable us to specify exactly the set of n for which Π_n is constructible. The key to the specification is the result (Theorems 4.8 and 11.3) that a geometric construction is possible if and only if the degree of the associated field extension is a power of 2.

Note first that the construction of Π_n depends on the construction of the angle $\theta_n = 2\pi/n$ at the centre of the polygon, for once we construct the isosceles triangle IOA for which the angle IOA is θ_n, we may form the polygon by pasting copies of the triangle all the way round.

A similar pasting technique allows us to deduce that constructibility of θ_m and θ_n implies constructibility of $\theta_m \pm \theta_n$:

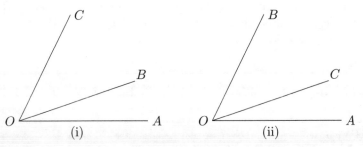

In both diagrams, AOB is the angle θ_m. In (i), $\angle BOC = \theta_n$ and $\angle AOC = \theta_m + \theta_n$. In (ii), $\angle COB = \theta_n$ and $\angle AOC = \theta_m - \theta_n$.

More generally, by repeated additions and subtractions, we obtain the following:

Theorem 11.5

If θ_m and θ_n are constructible, and if $s, t \in \mathbb{Z}$, then $s\theta_m + t\theta_n$ is constructible.

We shall have occasion to use the following fairly obvious remark concerning the constructibility of angles and points.

Theorem 11.6

The following statements are equivalent:

(i) θ_n is constructible;

(ii) the point $(\cos \theta_n, \sin \theta_n)$ is constructible;

(iii) the point $(\cos \theta_n, 0)$ is constructible.

Proof

(i) $\Rightarrow$ (ii). This is clear from the diagram.

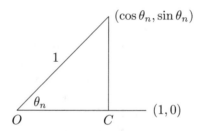

(ii) $\Rightarrow$ (iii). This is clear from Lemma 11.1.

(iii) $\Rightarrow$ (i). In the diagram, if we have constructed the point $C(\cos \theta_n, 0)$, then the line though C perpendicular to OI meets the circle with centre O and radius 1 in the point $(\cos \theta_n, \sin \theta_n)$. Joining this point to O gives the required angle. $\qquad \square$

The following lemma plays a crucial part in the proof of the main theorem:

Lemma 11.7

Let m and n be relatively prime positive integers. Then Π_{mn} is constructible if and only if Π_m and Π_n are constructible.

Proof

Suppose first that Π_{mn}, with vertices

$$V_0, V_1, \ldots V_{mn-1},$$

is constructible. It is clear that Π_m is constructible: simply join up the vertices $V_0, V_n, V_{2n}, \ldots, V_{(m-1)n}, V_0$ in sequence. Similarly, Π_n is constructible. (We have not used "relatively prime" in this part of the proof.)

Conversely, suppose that Π_m and Π_n are constructible, where m and n are relatively prime. Then there exist integers s and t such that $sm + tn = 1$, and so

$$s\theta_n + t\theta_m = \frac{2\pi s}{n} + \frac{2\pi t}{m} = \frac{2\pi(sm + tn)}{mn} = \theta_{mn}.$$

By Theorem 11.5, $s\theta_n + t\theta_m$ is constructible, and so θ_{mn} is constructible. $\square$

The following lemma will shortly be useful:

Lemma 11.8

Let $\omega_p = e^{\theta_p}(= e^{2\pi i/p})$, where p is prime. Then θ_p is constructible if and only if $[\mathbb{Q}(\omega_p) : \mathbb{Q}]$ is a power of 2.

Proof

Let $\omega = e^{2\pi i/p}$. Over the field $\mathbb{Q}(\omega)$ the polynomial $X^p - 1$ factorises as

$$(X - 1)(X - \omega)(X - \omega^2)\ldots(X - \omega^{p-1}),$$

and it follows that $\mathbb{Q}(\omega)$ is the splitting field over $\mathbb{Q}$ of the polynomial

$$\frac{X^p - 1}{X - 1} = X^{p-1} + X^{p-2} + \cdots + X + 1.$$

The polynomial is irreducible over $\mathbb{Q}$, by Example 2.31, and, by Corollary 8.14, $\text{Gal}(\mathbb{Q}(\omega) : \mathbb{Q})$ is abelian. Let $K = \mathbb{Q}(\omega) \cap \mathbb{R}$, a subfield of $\mathbb{R}$ containing $\zeta = (\omega + \omega^{-1})/2 = \cos(2\pi/p)$. The minimum polynomial of ω over K is $X^2 - 2\zeta X + 1$, and so $[\mathbb{Q}(\omega) : K] = 2$. Hence $\text{Gal}(\mathbb{Q}(\omega) : K)$ is a subgroup of $\text{Gal}(\mathbb{Q}(\omega) : \mathbb{Q})$ of order 2. It is certainly a normal subgroup, since $\text{Gal}(\mathbb{Q}(\omega) : \mathbb{Q})$ is abelian. Hence, by Theorem 7.34, the extension $K : \mathbb{Q}$ is normal. By Theorems 11.4 and 4.8, $2\pi/p$ is constructible if and only if $[K : \mathbb{Q}]$ is a power of 2, and hence (since $[\mathbb{Q}(\omega) : \mathbb{Q}] = 2[K : \mathbb{Q}]$) if and only if $[\mathbb{Q}(\omega) : \mathbb{Q}]$ is a power of 2. $\square$

We are ready now to prove the main theorem of this chapter:

Theorem 11.9

A regular polygon with n sides is constructible if and only if $n = 2^k p_1 p_2 \ldots p_r$, where k and r are non-negative integers and $p_1, p_2, \ldots, p_r$ are distinct prime numbers of the form $2^{2^m} + 1$.

Proof

Let
$$n = p_1^{m_1} p_2^{m_2} \ldots p_s^{m_s},$$
where $p_1, p_2, \ldots, p_s$ are distinct primes and $m_1, m_2, \ldots, m_s \geq 1$, and suppose that Π_n is constructible. By Lemma 11.7, Π_q is constructible, where $q = p^m$ is any one of the factors $p_j^{m_j}$. It follows that the point $(\cos \theta_q, \sin \theta_q)$ is constructible, where, as usual, we are writing $\theta_q = 2\pi/q$. Hence, by Theorem 4.8, $[\mathbb{Q}(\cos \theta_q, \sin \theta_q) : \mathbb{Q}]$ is a power of 2. Since $\mathbb{Q}(\omega) = \mathbb{Q}(\cos \theta_q, \sin \theta_q, i)$ is an extension of $\mathbb{Q}(\cos \theta_q, \sin \theta_q)$ of degree 2, it follows that $[\mathbb{Q}(\omega) : \mathbb{Q}]$ is also a power of 2.

The complex number ω is a **primitive** qth root of unity. That is,
$$\omega^{p^m} = 1, \text{ but } \omega^{p^{m-1}} \neq 1. \tag{11.1}$$

Lemma 11.10

The minimum polynomial of ω is
$$f = 1 + X^{p^{m-1}} + X^{2p^{m-1}} + \cdots + X^{(p-1)p^{m-1}}.$$

Proof

Writing $X^{p^{m-1}}$ as Z, we easily see that
$$f = 1 + Z + \cdots + Z^{p-1} = \frac{Z^p - 1}{Z - 1} = \frac{X^{p^m} - 1}{X^{p^{m-1}} - 1},$$
and from (11.1) we see that $f(\omega) = 0$. It remains to show that f is irreducible over $\mathbb{Q}$. Let $X = 1 + T$; then
$$f = \frac{(1 + T)^{p^m} - 1}{(1 + T)^{p^{m-1}} - 1}.$$
All the intermediate binomial coefficients are divisible by p, and so we may write
$$f = \frac{T^{p^m} + pu(T)}{T^{p^{m-1}} + pv(T)},$$
where u and v are polynomials, and $\partial u \leq p^m - 1$, $\partial v \leq p^{m-1} - 1$. Hence
$$f = T^{p^{m-1}(p-1)} + \frac{pu(t) - pT^{p^{m-1}(p-1)}v(T)}{T^{p^{m-1}} + pv(T)}. \tag{11.2}$$

The degree of the numerator $pu(t) - pT^{p^{m-1}(p-1)}$ is less than p^m and the degree of the denominator is p^{m-1}. Since f is a polynomial in T, the fractional term in (11.2) must be a polynomial of degree less than $p^{m-1}(p-1)$. Moreover, since the numerator is divisible by p and the denominator is not, we may write

$$f = T^{p^{m-1}(p-1)} + pg(T),\qquad(11.3)$$

where g is a polynomial and $\partial g < p^{m-1}(p-1)$.

We have an alternative expression

$$f(1+T) = 1 + (1+T)^{p^{m-1}} + (1+T)^{2p^{m-1}} + \cdots + (1+T)^{(p-1)p^{m-1}},$$

and from this it is evident that the constant term of $f(1+T)$ is p. From (11.3) it now follows by the Eisenstein criterion (Theorem 2.27) that f is irreducible. $\qquad\square$

Returning to the proof of Theorem 11.9, we now reconcile the two bits of information we have on $q = p^m$. On the one hand, we know that $[\mathbb{Q}(e^{2\pi i/q} : \mathbb{Q}] = 2^r$, a power of 2; and, on the other hand, we know that $[\mathbb{Q}(e^{2\pi i/q} : \mathbb{Q}] = p^{m-1}(p-1)$. If $p = 2$, there is no conflict at all between these statements; but if p is odd, we are forced to conclude that $m = 1$ and that $p-1$ is a power of 2. Suppose that $p = 2^k + 1$, and suppose that $k = 2^v u$ where $u > 1$ is odd. Then, writing 2^{2^v} as w, we have that

$$p = w^u + 1 = (w+1)(w^{u-1} - w^{u-2} + \cdots - w + 1),$$

which is impossible, since p is prime. Hence k has no odd factors, and we conclude that p is a **Fermat**[1] **prime,** of the form $2^{2^m} + 1$. We have shown that if Π_n is constructible, then $n = 2^k p_1 p_2 \ldots p_r$, where each p_i is a Fermat prime.

Conversely, suppose that $n = 2^k p_1 p_2 \ldots p_r$, where each $p_j = 2^{2^{m_j}} + 1$ is a Fermat prime. It will follow that Π_n is constructible if Π_{2^k} and Π_{p_j} ($i = 1, 2, \ldots, r$) are constructible. From Exercise 4.2 we can repeatedly bisect the angle $\pi/2$ to obtain $\pi/2^{k-1}$, and so Π_{2^k} is constructible. We must show that each Π_{p_j} is constructible. Let $\omega = e^{2\pi i/p_j}$. Then, by Lemma 11.8, $\mathbb{Q}(\omega)$ is a normal extension of $\mathbb{Q}$, with $[\mathbb{Q}(\omega) : \mathbb{Q}] = p_j - 1 = 2^{2^{m_j}}$. By Lemma 11.3, the angle $2\pi i/p_j$ is constructible. $\qquad\square$

Remark 11.11

The only known Fermat primes $F_m = 2^{2^m} + 1$ are

m	0	1	2	3	4
F_m	3	5	17	257	65,537

---

[1] Pierre de Fermat, 1601–1665.

EXERCISE

11.1. List the numbers n between 3 and 100 for which Π_n is constructible.

12
Solutions

Chapter 1

1.1.　(i) From (R3), $0 = 0 + 0$. Hence, from (R6), $a0 = a(0+0) = a0 + a0$, and so, from (R4), (R1) and (R3), $0 = a0 + \left(- (a0) \right) = (a0 + a0) + \left(- (a0) \right) = a0 + \left[a0 + \left(- (a0) \right) \right] = a0 + 0 = a0$.

(ii) From (R6), (R4) and (i), $ab + a(-b) = a \left(b + (-b) \right) = a0 = 0$. Hence, from (R3), (R1) and (R2), $-ab = -ab + 0 = -ab + \left(ab + a(-b) \right) = (-ab + ab) + a(-b) = 0 + a(-b) = a(-b)$. The proof of $(-a)b = -ab$ is similar.

First, for all a in R, we have $-(-a) = a$; for from (R3), (R4), (R1) and (R2), $a = a + 0 = a + \left[-a + \left(- (-a) \right) \right] = \left(a + (-a) \right) + \left(- (-a) \right) = 0 + \left(- (-a) \right) = -(-a)$. Substitute $-a$ for a in the identity $a(-b) = -ab$ to obtain $(-a)(-b) = - \left((-a)b \right) = - \left(- (ab) \right) = ab$.

1.2.　If $1 = 0$ then, for all a in R, $a = a1 = a0 = 0$, and so $R = \{0\}$.

1.3.　Let $K = \{0, 1\}$. As far as addition is concerned, the property of 0 makes it clear that $0 + 0 = 0$, $1 + 0 = 0 + 1 = 1$, and the only thing left for $1 + 1$ is 0. We obtain the cyclic group of order 2, and so axioms (R1), (R2), (R3) and (R4) are satisfied. The multiplication table must be

	0	1
0	0	0
1	0	1

For Axiom (R5), it is clear that $(ab)c = a(bc) = 1$ if $a = b = c = 1$, and otherwise $(ab)c = a(bc) = 0$. As for (R7), it is clear that $ab = ba = 1$ if $a = b = 1$, and otherwise $ab = ba = 0$. Since multiplication is commutative, we need only verify the first of the distributive laws: if $a = 0$, then $a(b + c) = ab + ac = 0$, and if $a = 1$, then $a(b+c) = ab+ac = b+c$. Thus K is a commutative ring with unity, and is indeed a field, since $1^{-1} = 1$.

1.4. Suppose that we have (R9), and let $ab = 0$ and $a \neq 0$. Then $ab = a0$ and so $b = 0$ by cancellation. Conversely, suppose that we have (R9)$'$, and let $ca = cb$, with $c \neq 0$. Then $c\Big(a + (-b)\Big) = ca + c(-b) = cb + c(-b) = cb + (-cb) = 0$, and so $a + (-b) = 0$. Hence $a = a + 0 = a + \Big((-b) + b\Big) = \Big(a + (-b)\Big) + b = 0 + b = b$.

1.5. Suppose that D is a finite integral domain, with elements $d_1, d_2, \ldots, d_n$, and let a ($= d_i$ for some i) be a non-zero element of D. The cancellation property (R9) has the consequence that the elements $ad_1, ad_2, \ldots, ad_n$ are all distinct, and so constitute a list of all the elements of D. Hence there must be some d_j in D with the property that $ad_j = 1$. Thus D is a field.

1.6. (i) $a = 1a$, and so $a \sim a$. (ii) If $a \sim b$, then $a = ub$, where u is in the group U of units. Hence $b = u^{-1}a$ and so $b \sim a$. (iii) If $a \sim b$ and $b \sim c$, then $a = ua$ and $b = vc$ for some u, v in U. Hence $a = (uv)c$, with $uv \in U$, and so $a \sim c$.

1.7. Suppose that $a + bi\sqrt{2}$ has an inverse $x + yi\sqrt{2}$. Then

$$|a + bi\sqrt{2}|^2 \, |x + yi\sqrt{2}|^2 = 1 \,.$$

That is, $(a^2 + 2b^2)(x^2 + 2y^2) = 1$. Since both the factors are positive integers, we must have that $a^2 + 2b^2 = 1$; and since a and b are integers, this can happen only if $a = \pm1$ and $b = 0$.

1.8. Since $a = 1a$, the property (i) is clear. If $b = xa$ and $c = yb$, with x, y in D, then $c = (yx)a$. Thus (ii) is proved – and so far we have used only the properties of a commutative ring. For (iii), suppose that $b = xa$ and $a = yb$. Then $1b = b = (xy)b$, and so, by cancellation, $xy = 1$. Hence x and y are units, and so $a \sim b$.

1.9. If $a + a = a$, then $a = a + 0 = a + \Big(a + (-a)\Big) = (a + a) + (-a) = a + (-a) = 0$.

1.10. Suppose that U satisfies (1.1). Then $0 = a - a \in U$ and so $-a = 0 - a \in U$ for all a in U. If $a, b \in U$, then $a - (-b) = a + b \in U$. Conversely, if U satisfies (1.2) and $a, b \in U$, then $-b \in U$ and so $a + (-b) = a - b \in U$.

1.11. Let U be a subring, as defined in the text. Then U contains at least one element a, and so $a - a = 0 \in U$. Since $0, a \in U$ we deduce that $0 - a = -a \in U$. If $a, b \in U$, then $a, -b \in U$, and so $a + b = a - (-b) \in U$ Thus U is closed with respect to the binary relations $+$ of addition and multiplication. The axioms (R1), (R2), (R5) and (R6) are automatic, and we have already shown that (R3) and (R4) hold. Hence U is a ring.

Conversely, suppose that U is a subset of the ring R, and is itself a ring with respect to the addition and multiplication in R. Thus, if $a, b \in U$, then $a+b, ab \in U$. Also U contains a zero element 0_U, which certainly has the property that $0_U + 0_U = 0_U$. Hence, by Exercise 1.9, $0_U = 0$. Within U, every element u has a negative u'. It also has a negative $-u$ within R, and $u + u' = u + (-u) = 0$. Hence $u' = 0 + u' = \Big((-u) + u\Big) + u' = (-u) + (u + u') = -u + 0 = -u$. Hence, if $a, b \in U$, it follows that $a - b \in U$.

1.12. Let E be a subfield of K, as defined in the text. Then K is certainly a ring with respect to the operations in K, by the previous exercise. Since E contains at least one non-zero element a, it follows that $1 = aa^{-1} \in E$. Consequently for *every* non-zero a in E, $a^{-1} = 1a^{-1} \in E$. Hence, for every a and every $b \neq 0$ in E, we have $ab = a(b^{-1})^{-1} \in K$. If $b = 0$ then, trivially, $ab = 0 \in E$. Thus E is closed under the addition and multiplication in E. Axioms (R1), (R2), (R3), (R4), (R5), (R6) and (R7) are satisfied, and we have already made sure of (R8) and (R10). Thus E is a field.

Conversely, suppose that E is a non-empty subset of K, and is a field with respect to the addition and multiplication in K. From the previous exercise, we know that E contains 0, that $-a \in E$ for every a in E, and that $a - b \in E$ for all a, b in E. There is a unity element 1_E in E, and from $1_E 1_E = 1_E$ we can deduce, by an argument essentially identical to the one used in the previous exercise, that 1_E coincides with the unity element 1 of K. Moreover, if $a \in E \setminus \{0\}$, its inverse a' in E coincides with its inverse a^{-1} in K, for $a' = (a^{-1}a)a' = a^{-1}(aa') = a^{-1}$. Hence, for all a and for all $b \neq 0$ in E, $ab^{-1} \in K$.

1.13. Suppose first that K is a field. Let $I \neq \{0\}$ be an ideal, and let a be a non-zero element of I. Then, for all b in K, we have $b = (ba^{-1})a \in I$, and so $I = K$.

Conversely, let K be a commutative ring with unity, having no proper ideals, and let $a \in K \setminus \{0\}$. Then $\langle a \rangle = K$, and so in particular there exists b in K such that $ba = 1$. Hence K is a field.

1.14. If $a + bi\sqrt{3}, c + di\sqrt{3} \in \mathbb{Q}(i\sqrt{3})$, then $(a + bi\sqrt{3}) - (c + di\sqrt{3}) = (a + c) - (b + d)i\sqrt{3} \in \mathbb{Q}(i\sqrt{3})$, and $(a + bi\sqrt{3})(c + di\sqrt{3}) = (ac - 3b) + (bc + ad)i\sqrt{3} \in \mathbb{Q}(i\sqrt{3})$. Also, if $c + di\sqrt{3} \neq 0$, then $|c + di\sqrt{2}|^2 = c^2 + 3d^2 \neq 0$, and

$$(c + di\sqrt{3})^{-1} = \frac{c}{c^2 + 3d^2} + \frac{-d}{c^2 + 3d^2}\, i\sqrt{3} \in \mathbb{Q}(i\sqrt{3}\,.$$

1.15. (i) K is a subring of the ring of all 2×2 matrices over $\mathbb{Q}$, since

$$\begin{pmatrix} a & b \\ -3b & a \end{pmatrix} - \begin{pmatrix} c & d \\ -3d & c \end{pmatrix} = \begin{pmatrix} a - c & b - d \\ -3(b - d) & a - c \end{pmatrix},$$

and

$$\begin{pmatrix} a & b \\ -3b & a \end{pmatrix}\begin{pmatrix} c & d \\ -3d & c \end{pmatrix} = \begin{pmatrix} ac - 3bd & ad + bc \\ -3(ad + bc) & ac - 3bd \end{pmatrix}.$$

Commutativity is easily verified, the multiplicative identity is obtained by putting $a = 1$ and $b = 0$, and

$$\begin{pmatrix} a & b \\ -3b & a \end{pmatrix}^{-1} = \frac{1}{a^2 + 3b^2}\begin{pmatrix} a & -b \\ 3b & a \end{pmatrix}.$$

(ii) The map

$$\begin{pmatrix} a & b \\ -3b & a \end{pmatrix} \mapsto a + bi\sqrt{3}$$

is an isomorphism.

1.16. The set $\mathbb{R}(i\sqrt{3})$ is a subfield of $\mathbb{C}$, and the argument is identical to that for $\mathbb{Q}(i\sqrt{3})$. The set $\mathbb{R}(\sqrt{3})$ is indeed a subfield of $\mathbb{R}$, since it coincides with $\mathbb{R}$. (Every real number x can be written as $0 + y\sqrt{3}$, where $y = x/\sqrt{3}$.)

1.17. Since $\ker \varphi$ is an ideal of K, it follows from the previous exercise that $\ker \varphi = \{0\}$. Hence, for all a, b in K,

$$\varphi(a) = \varphi(b) \Rightarrow a - b \in \ker \varphi = \{0\} \Rightarrow a = b.$$

1.18. In an integral domain we have the property that $a^2 = a \Rightarrow a = 0$ or $a = 1$; for, if $a \neq 0$, the cancellation property gives $aa = a = a1 \Rightarrow a = 1$. From $1_R 1_R = 1_R$ we deduce that $\varphi(1_R)\varphi(1_R) = \varphi(1_R)$ and so, since S is an integral domain, $\varphi(1_R) = 1_S$.

Let $R = \mathbb{Z}$ and S be the set of all matrices

$$\begin{pmatrix} a & 0 \\ 0 & b \end{pmatrix},$$

where $a, b \in \mathbb{Z}$. Under matrix addition and multiplication, S is a commutative ring, with the 2×2 identity matrix as unity element. It is not an integral domain, since, for example,

$$\begin{pmatrix} a & 0 \\ 0 & 0 \end{pmatrix} \begin{pmatrix} 0 & 0 \\ 0 & b \end{pmatrix} = \begin{pmatrix} 0 & 0 \\ 0 & 0 \end{pmatrix}.$$

Let $\varphi : R \to S$ be given by

$$\varphi(a) = \begin{pmatrix} a & 0 \\ 0 & 0 \end{pmatrix} \quad (a \in R).$$

Then $\varphi(1_R) \neq 1_S$.

1.19.
$$\left(\frac{a}{b} + \frac{c}{d} \right) + \frac{e}{f} = \frac{ad + bc}{bd} + \frac{e}{f} = \frac{(ad + bc)f + (bd)e}{bd(f)}$$
$$= \frac{a(df) + b(cf + de)}{b(df)} = \frac{a}{b} + \frac{cf + de}{df} = \frac{a}{b} + \left(\frac{c}{d} + \frac{d}{f} \right).$$

1.20. Suppose that D is a field. Then, if θ is the monomorphism featuring in Theorem 1.3,

$$\frac{a}{b} = \frac{ab^{-1}}{bb^{-1}} = \frac{ab^{-1}}{1} = \theta(ab^{-1}).$$

Thus θ is onto as well as one-one, and so $Q(D) \simeq D$.

1.21. Since $6(1_K) = 0_K$, and since no positive integer smaller than 6 has this property, $\mathrm{char}\,(\mathbb{Z}_6) = 6$. For all a in $\mathbb{Z}_6$, $a^2 = 0 \Rightarrow 6 \mid a^2 \Rightarrow 2 \mid a^2$ and $3 \mid a^2 \Rightarrow 2 \mid a$ and $3 \mid a \Rightarrow 6 \mid a \Rightarrow a = 0$.

This property holds for $\mathbb{Z}_n$ if and only if n is **square-free** (not divisible by the square of any prime number). The argument used for 6 works for any product $p_1 p_2 \ldots p_k$ of distinct primes. Conversely, if $n = p^2 m$ for some prime p, then, in $\mathbb{Z}_n$, $pm \neq 0$ but $(pm)^2 = 0$.

1.22. Leaving out 0, we have the table

	1	2	3	4	5	6
1	1	2	3	4	5	6
2	2	4	6	1	3	5
3	3	6	9	5	1	4
4	4	1	5	2	6	3
5	5	3	1	6	4	2
6	6	5	4	3	2	1

$$1^{-1} = 1,\, 2^{-1} = 4,\, 3^{-1} = 5,\, 4^{-1} = 2,\, 5^{-1} = 3,\, 6^{-1} = 6.$$

1.23. This is certainly true for $n = 1$, since

$$(a + b)^1 = a + b = \binom{1}{0}a + \binom{1}{1}b\,.$$

Let $n \geq 1$, and suppose inductively that the theorem is true for n. To show that it is true for $n + 1$ we require the Pascal identity

$$\binom{n}{r-1} + \binom{n}{r} = \binom{n+1}{r} \quad (n \geq r \geq 1), \tag{12.1}$$

whose proof is straightforward. For the induction step, during which we use only operations that are valid in a commutative ring, we have that

$$(a + b)^{n+1} = (a + b)(a + b)^n = (a + b)\left(\sum_{r=0}^{n}\binom{n}{r}a^{n-r}b^r\right)$$

$$= (a + b)\left(a^n + \cdots + \binom{n}{r-1}a^{n-r+1}b^{r-1} + \binom{n}{r}a^{n-r}b^r + \cdots + b^n\right).$$

For $r = 1, 2, \ldots, n$, the term in $a^{(n+1)-r}b^r$ is

$$b \cdot \binom{n}{r-1}a^{n-r+1}b^{r-1} + a \cdot \binom{n}{r}a^{n-r}b^r\,,$$

and so the coefficient is

$$\binom{n}{r-1} + \binom{n}{r} = \binom{n+1}{r}\,.$$

The coefficients of a^{n+1} and b^{n+1} are both 1, and so we conclude that

$$(a + b)^{n+1} = \sum_{r=0}^{n+1}\binom{n+1}{r}a^{(n+1)-r}b^r\,.$$

Hence, by induction, the result is true for all $n \geq 1$.

1.24. As in Theorem 1.17, we have that

$$(x - y)^p = \sum_{r=0}^{p}\binom{p}{r}(-1)^r x^{p-r}y^r = x^p + (-1)^p y^p\,.$$

If p is odd, the result is immediate. If $p = 2$ it seems that $(x - y)^2 = x^2 + y^2$, but $a = -a$ for every a in a field of characteristic 2, and so the result is still true.

1.25. It is true for $n = 1$. Suppose that $(x \pm y)^{p^{n-1}} = x^{p^{n-1}} \pm y^{p^{n-1}}$. Then

$$(x \pm y)^{p^n} = [(x \pm y)^{p^{n-1}}]^p = (x^{p^{n-1}} + y^{p^{n-1}})^p = x^{p^n} \pm y^{p^n}\,.$$

1.26. If H has index 2 in G, then the cosets, both left and right, must be H and $G \setminus H$. Hence H is normal.

1.27. The group $(\mathbb{Z}_n, +)$ consists of "powers" $1, 1+1, 1+1+1, \ldots$, and so is cyclic, generated by 1.

1.28. Let $G = \langle a \rangle$ and let H be a proper subgroup of G. Then $a \notin H$, and there exists a smallest positive integer m with the property that $a^m \in H$. If $a^n \in H$, then m divides n, for n can be written as $qm + r$, with $0 \le r \le m - 1$, and $a^r = a^n (a^m)^{-q} \in H$, a contradiction unless $r = 0$. Thus H is cyclic, generated by a^m.

1.29. Since $b^2 = q^2 = e$, both B and Q are subgroups. For B the cosets, both left and right, are B, $Ba = aB = \{a,c\}$, $Bp = pB = \{p,r\}$, $Bq = qB = \{q,s\}$. Thus B is normal. For Q, the left cosets are Q, $Qa = \{a,p\}$, $Qb = \{b,s\}$, $Qc = \{c,r\}$, and the right cosets are Q, $aQ = \{a,r\}$, $bQ = \{b,s\}$, $cQ = \{c,p\}$. Thus Q is not normal.
Define φ by

$$\varphi(e) = \varphi(b) = e\,, \ \varphi(a) = \varphi(c) = x\,, \ \varphi(p) = \varphi(r) = y \ \varphi(q) = \varphi(s) = z\,.$$

1.30. Let $o(a) = m$, $o(b) = n$. Since A is abelian, $(ab)^{mn} = (a^m)^n (b^n)^m = e$, and so $o(ab)$ divides mn. In the given group, $o(x) = o(y) = 2$, but $o(xy) = o(a) = 3$.

1.31. Let H be a subgroup of G/N, and let $K = \{x \in G : Nx \in H\}$. Then K is a subgroup of G, since $x, y \in K$ implies that $Nx, Ny \in H$ and so $(Nx)(Ny)^{-1} = N(xy^{-1}) \in H$. If $y \in N$, then $Ny = N \in H$ and so $y \in K$. Thus K is a subgroup of G containing N, and $H = K/N = \{Nx \in G/H : x \in K\}$. The subgroup H is normal in G/N if and only if, for all Nx in G and all Ny in H,

$$(Nx)^{-1}(Ny)(Nx) \in H$$

that is, if and only if $N(x^{-1}yx) \in H$, that is (for all x in G and all y in K)

$$x_{-1}yx \in K\,,$$

that is, if and only if K is normal in G.

Chapter 2

2.1. (i) $1218 = 846 + 372$, $846 = 2 \times 372 + 102$, $372 = 3 \times 102 + 66$, $102 = 66 + 36$, $66 = 36 + 30$, $36 = 30 + 6$, $30 = 5 \times 6$. The last non-zero remainder is 6, and this is the greatest common divisor. Also, $6 = 36 - 30 = 36 - (66 - 36) = 2 \times 36 - 66 = 2 \times (102 - 66) - 66 = 2 \times 102 - 3 \times 66 = 2 \times 102 - 3(372 - 3 \times 102) = 11 \times 102 - 3 \times 372 = 11 \times (846 - 2 \times 372) - 3 \times 372 = 11 \times 846 - 25 \times 372 = 11 \times 846 - 25(1218 - 846) = 36 \times 846 - 25 \times 1,218$.

(ii) $851 = 779 + 72$, $779 = 10 \times 72 + 59$, $72 = 59 + 13$, $59 = 4 \times 13 + 7$, $13 = 7 + 6$, $7 = 6 + 1$. The greatest common divisor is 1. Also, $1 = 7 - 6 = 7 - (13 - 7) = 2 \times 7 - 13 = 2(59 - 4 \times 13) - 13 = 2 \times 59 - 9 \times 13 = 2 \times 59 - 9(72 - 59) = 11 \times 59 - 9 \times 72 = 11 \times (779 - 10 \times 72) - 9 \times 72 = 11 \times 779 - 119 \times 72 = 11 \times 779 - 119(851 - 779) = 130 \times 779 - 119 \times 851$.

2.2. Let D be an integral domain. Then it is embeddable in its field of quotients, by the results of Section 1.3. If R is a commutative ring with unity which is not an integral domain, then there exist a, b in $\mathbb{R} \setminus \{0\}$ such that $ab = 0$. This remains true in any ring of which R is a subring, and so R cannot be embedded in a field.

2.3. (i) Since $a, b \in \Gamma$ implies that $a - b, ab \in \Gamma$, we know that Γ is a subring of $\mathbb{C}$. From the previous exercise, it must therefore be an integral domain.

(ii) $a = (u + iv)b = [u' + iv' + (u - u') + i(v - v')]b = qb + r$, where $r = [(u - u') + i(v - v')]b$. Now $r = a - qb$, where $a, q, b \in \Gamma$, and so $r \in \Gamma$. Also

$$\delta(r) = [(u - u')^2 + (v - v')^2]\, \delta(b) \leq \tfrac{1}{2}\, \delta(b) < \delta(b)\,,$$

and so Γ is a euclidean domain.

2.4. (i) If $\frac{r}{s}, \frac{u}{v} \in R$, then $p \nmid sv$, and so

$$\frac{r}{s} - \frac{u}{v} = \frac{rv - su}{sv} = \frac{x}{y}\,,$$

where x and y are obtained by dividing $rv - su$ and sv by their greatest common divisor. Certainly $x/y \in R$. A similar argument shows that

$$\left(\frac{r}{s}\right)\left(\frac{u}{v}\right) \in R.$$

(ii) A non-zero element r/s in R has an inverse s/r in R if and only if $p \nmid r$.

(iii) Let I be a non-zero ideal in R. Let $I_k = \{p^k r/s \in I : p \nmid r\}$. If $I_0 \neq \emptyset$, then I contains a unit, and so $I = R$. Otherwise, let k be the smallest integer such that $I_k \neq \emptyset$, and let $p^k r/s \in I_k$. If $p^l u/v$ is an arbitrary element of I, then $l \geq k$, and so

$$\frac{p^l u}{v} = \frac{p^{l-k}}{1} \cdot \frac{p^k r}{s} \cdot \frac{su}{rv}$$

and so $p^l u/v \in \langle p^k r/s \rangle$.

2.5. (i) Let $u + vi \in \Gamma$. Then $(u + vi)(x + yi) = 1$ is possible only if $|u + vi|^2 = u^2 + v^2 = 1$. Since u and v are integers, we must have $u = \pm 1, v = 0$, or $u = 0, v = \pm 1$. Thus the group of units is $\{1, -1, i, -i\}$.

(ii) The number 5, while irreducible in $\mathbb{Z}$, factorises in Γ as $(1+2i)(1-2i)$. If we suppose that $(1+2i)$ factorises into $(a+bi)(c+di)$, then $(a^2 + b^2)(c^2 + d^2) = |1 + 2i|^2 = 5$. Hence one or other of $a^2 + b^2$ and $c^2 + d^2$ is equal to 1, and so either $a + bi$ or $c + di$ is a unit. Thus $1 + 2i$ (and similarly) $1 - 2i$ is irreducible.

(iii) No, for $3 + 2i = i(2 - 3i)$ and $3 - 2i = -i(2 + 3i)$. Hence $3 + 2i \sim 2 - 3i$ and $3 - 2i \sim 2 + 3i$.

2.6. (i) If $u = a + bi\sqrt{3}$ and $v = c + di\sqrt{3}$, then $\varphi(uv) = (ac - 3bd)^2 + 3(ad + bc)^2 = a^2 c^2 + 9b^2 d^2 + 3a^2 d^2 + 3b^2 c^2 = (a^2 + 3b^2)(c^2 + 3d^2) = \varphi(u)\varphi(v)$. It is clear that $\varphi(0) = 0$ and $\varphi(\pm 1) = 1$. Otherwise $\varphi(u) > 2$.

(ii) With u and v as above, if $uv = 1$, then $\varphi(u)\varphi(v) = \varphi(1) = 1$. This can happen only if $u = \pm 1$.

(iii) If $1 \pm i\sqrt{3}$ factorises non-trivially as $(a + bi\sqrt{3})(c + di\sqrt{3})$, then, applying φ, we obtain $4 = (a^2 + 3b^2)(c^2 + 3d^2)$. But each of the factors on the right is greater than 2, and we have contradiction.

(iv) The integer $2 = 2 + 0i\sqrt{3}$ is also irreducible. We thus have that $4 = 2 \times 2 = (1 + i\sqrt{3})(1 - i\sqrt{3})$, and so R is not a unique factorisation domain.

2.7. Let $f = a_0 + a_1 X + \cdots$, $g = b_0 + b_1 X + \cdots$, $h = c_0 + c_1 + \cdots$. The coefficient of X^k in $f(g + h)$ is

$$\sum_{\{(i,j)\,:\,i+j=k\}} a_i(b_j + c_j) = \sum_{\{(i,j)\,:\,i+j=k\}} a_i b_j + \sum_{\{(i,j)\,:\,i+j=k\}} a_i c_j)$$

and this is the coefficient of X^k in $fg + fh$.

2.8. (i) $X^3 + X + 1 = (X - 1)(X^2 + X + 1) + (X + 2)$.
       (ii) $X^7 + 1 = (X^4 - X)(X^3 + 1) + (X + 1)$.

2.9. Consider, for example, the ideal

$$I = \langle 2, X \rangle = \{2f(X) + Xg(X) \, : \, f(x), g(X) \in \mathbb{Z}[X]\},$$

consisting of all polynomials whose constant term is even. Suppose that $I = \langle p \rangle$ for some polynomial p. Then $p \mid 2$ and $p \mid X$, and so $p \sim 1$. But then $\langle p \rangle = \mathbb{Z}[X] \neq I$.

2.10. Consider, for example, the ideal $I = \langle X^2, Y^2 \rangle$, and suppose that $I = \langle f \rangle$ for some f in $K[X, Y]$. Then $f \mid X^2$ and $f \mid Y^2$, and so $f \sim 1$ and $\langle f \rangle = K[X, Y]$. But $I \neq K[X, Y]$, since, for example, $X \notin I$.

2.11. (i) First, $X^5 + X^4 - 2X^3 - X^2 + X = (X^2 + X - 3)(X^3 + X - 2) + (6X - 6)$. Next, $X^3 + X - 2 = (\frac{1}{6}X^2 + \frac{1}{6}X + \frac{1}{3})(6X - 6) + 0$. The greatest common divisor is $6X - 6) \sim X - 1$, the last non-zero remainder, and $6X - 6 = f - (X^2 + X - 3)g$.
       (ii) First, $X^3 + 2X^2 + 7X - 1 = (X - 1)(X^2 + 3X + 4) + (6X + 3)$. Next, $X^2 + 3X + 4 = (\frac{1}{6}X + \frac{5}{12})(6X + 3) + \frac{11}{4}$. The greatest common divisor is $\frac{11}{4}$ (~ 1), and

$$\begin{aligned}
\tfrac{11}{4} &= X^2 + 3X + 4 - (\tfrac{1}{6}X + \tfrac{5}{12})(6X + 3) \\
&= X^2 + 3X + 4 - (\tfrac{1}{6}X + \tfrac{5}{12})[X^3 + 2X^2 + 7X - 1 \\
&\quad - (X - 1)(X^2 + 3X + 4)] \\
&= [1 + (\tfrac{1}{6}X + \tfrac{5}{12})(X - 1)](X^2 + 3X + 4) \\
&\quad - (\tfrac{1}{6}X + \tfrac{5}{12})(X^3 + 2X^2 + 7X - 1) \\
&= (\tfrac{1}{6}X^2 + \tfrac{1}{4}X + \tfrac{7}{12})(X^2 + 3X + 4) \\
&\quad - (\tfrac{1}{6}X + \tfrac{5}{12})(X^3 + 2X^2 + 7X - 1).
\end{aligned}$$

2.12. The group $\mathbb{Z}_p^*$ of non-zero elements of $\mathbb{Z}_p$ is of order $p - 1$, and so $a^{p-1} = 1$ for all a in $\mathbb{Z}_p^*$. It follows that every element of $\mathbb{Z}_p$ (including 0) is a root of the the polynomial $X^p - X$. Thus, by Theorem 2.18, $X^p - X$ is divisible by $X(X-1)(X-2)\ldots(X-(p-1))$. Since this divisor, like $X^p - X$ itself, is monic and of degree p, the two polynomials must be equal.

2.13. Suppose that $f - g \neq 0$¿ Then $f - g$, of degree not greater than n, is divisible by $(X - \alpha_1)(X - \alpha_2)\ldots(X - \alpha_{n+1})$. This is impossible, and so $f - g = 0$.

2.14. By Gauss's lemma, if this factorises over $\mathbb{Q}$ then it factorises over $\mathbb{Z}$. One of the factors must be a linear factor $X - \alpha$, and $\alpha = \pm 1$ or ± 5. Since none of these four numbers is a root of the polynomial, it follows that no factorisation is possible.

2.15. These are all irreducible by the Eisenstein criterion, with $p = 3$, $p = 2$ and $p = 5$.

2.16. Let $Y = 1/X$. Then $5X^4 - 10X^3 + 10X - 3 = (-1/Y^4)(3Y^4 - 10Y^3 + 10Y - 5)$. Any non-trivial factorisation of $5X^4 - 10X^3 + 10X - 3$ would force a non-trivial factorisation of $3Y^4 - 10Y^3 + 10Y - 5$, and, by the Eisenstein criterion, this cannot happen.
$X^4 + 4X^3 + 3X^2 - 2X + 4 = (X^4 + 4X^3 + 6X^2 + 4X + 1) - 3X^2 - 6X + 3 = (X + 1)^4 - 3(X + 1)^2 + 6 = Y^4 - 3Y^2 + 6$, where $Y = X + 1$. Any non-trivial factorisation of $X^4 + 4X^3 + 3X^2 - 2X + 4$ would force a factorisation of $Y^4 - 3Y^2 + 6$, and, by the Eisenstein criterion, this cannot happen.

2.17. Let $g = 4X^4 - 2X^2 + X - 5$. The corresponding polynomial in $\mathbb{Z}_3[X]$ is $\bar{g} = X^4 + X^2 + X + 1$. This has no linear factors, since $\bar{g}(0) = 1$, $\bar{g}(1) = 1$ and $\bar{g}(-1) = -1$. Suppose that

$$X^4 + X^2 + X + 1 = (X^2 + aX + b)(X^2 + cX + d).$$

Then

$$a + c = 0 \ \text{(i)}, \quad ac + b + d = 1 \ \text{(ii)}, \quad ad + bc = 1 \ \text{(iii)}, \quad bd = 1 \ \text{(iv)}.$$

From (iv), either $b = d = 1$ or $b = d = -1$. In the former case (iii) becomes $a + c = 1$ and contradicts (i). In the latter case (iii) becomes $a + c = -1$, again a contradiction. Thus $\bar{g}$ is irreducible over $\mathbb{Z}_3$. Now any non-trivial factorisation of g over $\mathbb{Q}$ would translate into a factorisation of $\bar{g}$ over $\mathbb{Z}_3$, and we have shown that this cannot happen. Thus g is irreducible over $\mathbb{Q}$.

Now let $q = 3X^4 - 7X + 5$, and let $\bar{q} = X^4 + X + 1$ be the corresponding polynomial in $\mathbb{Z}_2[X]$. This has no linear factor, since $\bar{q}(0) = \bar{q}(1) = 1$. If

$$X^4 + X + 1 = (X^2 + aX + b)(X^2 + cX + d),$$

then

$$a + c = 0 \ \text{(i)}, \quad ac + b + d = 0 \ \text{(ii)}, \quad ad + bc = 1 \ \text{(iii)}, \quad bd = 1 \ \text{(iv)}.$$

From (iv) we must have $b = d = 1$, and so (iii) becomes $a + c = 1$, and contradicts (i). Thus $\bar{q}$, and hence also q, is irreducible.

Chapter 3

3.1.   (i) Since $[M : K] = [M : L][L : K]$, it follows from $[M : K] = [L : K]$ that $[M : L] = 1$. Thus $M = L$.
       (ii) Similarly, it follows from $[M : L] = [M : K]$ that $[L : K] = 1$, and so $L = K$.

3.2.   Since $[L : K] = [L : E][E : K]$ and $[L : K]$ is prime, either $[L : E] = 1$ or $[E : K] = 1$. Thus either $E = L$ or $E = K$.

3.3.   Let
$$M(a,b) = \begin{pmatrix} a & b \\ nb & a \end{pmatrix}.$$

Define $\phi : F \to \mathbb{Q}[\sqrt{n}]$ by $\phi\big(M(a,b)\big) = a + b\sqrt{n}$. Then ϕ clearly maps onto $\mathbb{Q}[\sqrt{n}]$, and it is one-to-one, since

$$a + b\sqrt{n} = a' + b'\sqrt{n} \Rightarrow a = a' \text{ and } b = b'. \qquad (12.2)$$

Next, $\phi\big(M(a,b) + M(c,d)\big) = \phi\big(M(a+c, b+d)\big) = (a+c) + (b+d)\sqrt{n} = \phi\big(M(a,b)\big) + \phi\big(M(c,d)\big)$, and $\phi\big(M(a,b)\,M(c,d)\big) = \phi\big(M(ac+nbd, ad+bc)\big) = (ac+nbd) + (ad+bc)\sqrt{n} = (a+b\sqrt{n})(c+d\sqrt{n}) = \phi\big(M(a,b)\big)\,\phi\big(M(c,d)\big)$. Thus ϕ is an isomorphism.

If n is a perfect square, the implication (12.2) fails, and so ϕ is not one-to-one.

3.4.   If $b = 0$, the minimum polynomial is $X - a$. So suppose that $b \neq 0$. Since $\big[\mathbb{Q}[\sqrt{2}] : \mathbb{Q}\big] = 2$, the minimum polynomial of $a + b\sqrt{2}$ must be of degree 2. Since $(a+b\sqrt{2})^2 = a^2 + 2b^2 + 2ab\sqrt{2} = -a^2 + 2b^2 + 2a(a+b\sqrt{2})$, the minimum polynomial is $X^2 - 2aX + (a^2 - 2b^2)$.

3.5.   If $\beta \in L \backslash K$, then $[K(\beta) : K] \geq 2$. Since $K(\beta) \subseteq L$ we must have $[K(\beta) : K] = 2$. By Theorem 3.3, $K(\beta) = L$. Since β is algebraic over K, it must have a minimum polynomial, and the only possible degree is 2.

3.6. $\alpha^3 + \alpha - 2 = \alpha(\alpha^2 + 2\alpha + 5) - 2(\alpha^2 + 2\alpha + 5) + 8 = 8$, and $\alpha^2 - 3 = (\alpha^2 + 2\alpha + 5) - 2\alpha - 8 = -2\alpha - 8$. So

$$\frac{\alpha^3 + \alpha - 2}{\alpha^2 - 3} = -\frac{4}{\alpha + 4}.$$

Next, dividing $X^2 + 2X + 5$ by $X + 4$ gives $X^2 + 2X + 5 = (X+4)(X-2) + 13$, and so $(\alpha + 4)(\alpha - 2) = -13$. Thus

$$\frac{\alpha^3 + \alpha - 2}{\alpha^2 - 3} = -\frac{4}{\alpha + 4} = \frac{4}{13}(\alpha - 2).$$

3.7.   Since $1 = -\alpha^3 - \alpha$, it is clear that $1/\alpha = (-\alpha^3 - \alpha)/\alpha = -\alpha^2 - 1$. Also, $(\alpha + 2)(\alpha^2 + p\alpha + q) = \alpha^3 + \alpha + r$ if and only if $p = -2$, $q = 5$ and $r = 10$. Thus $(\alpha + 2)(\alpha^2 - 2\alpha + 5) = (\alpha^3 + \alpha + 1) + 9 = 9$, and so $1/(\alpha + 2) = \frac{1}{9}(\alpha^2 - 2\alpha + 5)$.

3.8.   (i) Suppose, for a contradiction, that there exist a, b in $\mathbb{Q}$ such that $\sqrt{3} = a + b\sqrt{2}$, where b must be non-zero, since $\sqrt{3}$ is irrational. Then $a^2 = (\sqrt{3} - b\sqrt{2})^2 = (3 + 2b^2) - 2b\sqrt{6}$, and so $\sqrt{6} = (2b^2 - a^2 + 3)/2b \in \mathbb{Q}$. This is a contradiction.

(ii) $(\sqrt{2}+\sqrt{3})^2 = 5 + 2\sqrt{6} = 2\sqrt{3}(\sqrt{2}+\sqrt{3}) - 1$. Hence the minimum polynomial of $\sqrt{2} + \sqrt{3}$ over $\mathbb{Q}[\sqrt{3}]$ is $X^2 - 2\sqrt{3}X + 1$.

3.9.   Certainly $\mathbb{Q}[\sqrt{2} + \sqrt{5}] \subseteq \mathbb{Q}(\sqrt{2}, \sqrt{5})$. Conversely, observe that $(\sqrt{2}+\sqrt{5})^3 = 2\sqrt{2} + 6\sqrt{5} + 15\sqrt{2} + 5\sqrt{5} = 17\sqrt{2} + 11\sqrt{5}$. Since both $\sqrt{2}+\sqrt{5}$ and $17\sqrt{2}+11\sqrt{5}$ are in $\mathbb{Q}[\sqrt{2} + \sqrt{5}]$, it follows that $\sqrt{2}, \sqrt{5} \in \mathbb{Q}[\sqrt{2} + \sqrt{5}]$. Hence $\mathbb{Q}(\sqrt{2}, \sqrt{5}) \subseteq \mathbb{Q}[\sqrt{2} + \sqrt{5}]$.

Since $(\sqrt{2} + \sqrt{5})^4 = (7 + 2\sqrt{10})^2 = 89 + 28\sqrt{10}$, we see that $(\sqrt{2} + \sqrt{5})^4 -$ $14(\sqrt{2} + \sqrt{5})^2 + 9 = 0$, and the minimum polynomial over $\mathbb{Q}$ is $X^4 - 14X + 9$. Since $(\sqrt{2} + \sqrt{5})^2 = 7 + 2\sqrt{10} = 2\sqrt{2}(\sqrt{2} + \sqrt{5}) + 3$, the minimum polynomial over $\mathbb{Q}[\sqrt{2}]$ is $X^2 - 2\sqrt{2}X - 3$. Since $(\sqrt{2}+\sqrt{5})^2 = 7+2\sqrt{10} = 2\sqrt{5}(\sqrt{2}+\sqrt{5})-3$, the minimum polynomial over $\mathbb{Q}[\sqrt{5}]$ is $X^2 - 2\sqrt{5}X + 3$.

3.10. The element $1 + \sqrt{3} \in \mathbb{Q}[\sqrt{3}] \setminus \mathbb{Q}$, and so it has minimum polynomial of degree 2. Since $(1 + \sqrt{3})^2 = 4 + 2\sqrt{3} = 2(1 + \sqrt{3}) + 2$, it follows that the minimum polynomial is $X^2 - 2X - 2$.
The element $\sqrt{3}/\sqrt{5}$ lies in $\mathbb{Q}(\sqrt{3}, \sqrt{5}) \setminus \mathbb{Q}$ and so has minimum polynomials of degree 2 or 4. Since $(\sqrt{3}/\sqrt{5})^2 = 3/5$, the minimum polynomial is $X^2 - (3/5)$. Since $(\sqrt{3}+\sqrt{5})^2 = 8+2\sqrt{15}$ and $(\sqrt{3}+\sqrt{5})^4 = 124+32\sqrt{15} = 16(8+2\sqrt{15})-4$, the minimum polynomial is $X^4 - 16X^2 + 4$.
The element $(1+i)\sqrt{3}$ lies in $\mathbb{Q}(i, \sqrt{3})$ and is in not in $\mathbb{Q}[\sqrt{3}]$, $\mathbb{Q}[i]$ or $\mathbb{Q}[i\sqrt{3}]$. So its minimum polynomial is of degree 4. Since $[(1+i)\sqrt{3}]^2 = 6i$ and $[(1+i)\sqrt{3}]^4 = -36$, the minimum polynomial is $X^4 + 36$.

3.11. Let $\alpha = \sqrt{1 + \sqrt{2}}$. Then $\alpha^2 = 1 + \sqrt{2}$ and so the minimum polynomial over $\mathbb{Q}[\sqrt{2}]$ is $X^2 - (1 + \sqrt{2})$. Since $(\alpha^2 - 1)^2 = 2$, the minimum polynomial over $\mathbb{Q}$ is $X^4 - 2X^2 - 1$.

3.12. $(1 + \sqrt{2} + \sqrt{3} + \sqrt{6})(a + b\sqrt{2} + c\sqrt{3} + d\sqrt{6}) = 1$ if and only if

$$a + 2b + 3c + 6d = 1$$
$$a + b + 3c + 3d = 0$$
$$a + c + 2b + 2d = 0$$
$$a + b + c + d = 0.$$

Solving these equations gives $a = d = 1/2$, $b = c = -1/2$. So the inverse is $(1/2)(1 - \sqrt{2} - \sqrt{3} + \sqrt{6})$.

3.13. The two statements are in fact the same. If g factorises non-trivially over K, so that $g = uv$, with $0 < \partial u < \partial p$ and $0 < \partial v < \partial p$, then the factors u and v are certainly in $L[x]$, and so g factorises also over L. Consequently, if g does *not* factorise over L, it does not factorise over K.

3.14. We have seen that the field $\mathbb{A}$ of algebraic numbers is countable. Hence certainly $\mathbb{R} \cap \mathbb{A}$, the field of real algebraic numbers, is countable. Since $\mathbb{R}$ is uncountable, there are $2^{\aleph_0}$ real transcendental numbers.

3.15. (i) This is true, since both $\mathbb{Q}(\alpha)$ and $\mathbb{Q}(\beta)$ are isomorphic to the field $\mathbb{Q}(X)$ of rational forms over $\mathbb{Q}$.
(ii) This is false. Let α be transcendental. If $1/\alpha$ were algebraic, with minimum polynomial $X^n + a_{n-1}X^{n-1} + \cdots + a_1 X + a_0$, then $(1/\alpha^n)(1 + a_{n-1}\alpha + \cdots + a_1\alpha^{n-1} + a_0\alpha^n) = 0$, and it would follow that α is algebraic. Thus $1/\alpha$ is transcendental. Taking β as $1/\alpha$, we see that the product of two transcendental numbers need not be transcendental.
(iii) This is false. Let $\alpha = e$ and $\beta = i\pi$. Then $\alpha^\beta = -1$.
(iv) This is true. If α^2 were algebraic, there would exist $a_0, a_1, \ldots, a_n$ (not all zero) such that $a_0 + a_1\alpha^2 + \cdots + a_n\alpha^{2n} = 0$, and this would immediately imply that α is algebraic.

3.16. (i) Expanding the determinant Δ_n by the first row, and using the induction hypothesis, we see that

$$\Delta_n = \lambda\Delta_{n-1} + (-1)^{n-1}q_n \begin{vmatrix} -1 & \lambda & 0 & \cdots & 0 \\ 0 & -1 & \lambda & \cdots & 0 \\ 0 & \vdots & \ddots & \ddots & \vdots \\ 0 & 0 & \cdots & -1 & \lambda \\ 0 & 0 & \cdots & \cdots & -1 \end{vmatrix}$$

$$= \lambda(q_{n-1} + q_{n-2}\lambda + \cdots + q_1\lambda^{n-2} + \lambda^{n-1}) + q_n$$

$$= q_n + q_{n-1}\lambda + \cdots + q_1\lambda^{n-1} + \lambda^n .$$

(ii) The matrix of T_α is

$$A = \begin{bmatrix} 0 & 0 & 0 & 0 & \cdots & -a_0 \\ 1 & 0 & 0 & 0 & \cdots & -a_1 \\ 0 & 1 & 0 & 0 & \cdots & -a_2 \\ 0 & 0 & 1 & 0 & \cdots & -a_3 \\ \vdots & \vdots & \vdots & \ddots & \ddots & \vdots \\ 0 & 0 & 0 & \cdots & 1 & -a_{n-1} \end{bmatrix},$$

and

$$|XI_n - A| = \begin{vmatrix} X & 0 & 0 & 0 & \cdots & a_0 \\ -1 & X & 0 & 0 & \cdots & a_1 \\ 0 & -1 & X & 0 & \cdots & a_2 \\ 0 & 0 & -1 & X & \cdots & a_3 \\ \vdots & \vdots & \vdots & \ddots & \ddots & \vdots \\ 0 & 0 & 0 & \cdots & -1 & X + a_{n-1} \end{vmatrix}.$$

By part (i), this is equal to $m(X)$.

3.17. Since $\beta = \alpha^2$ does not belong to K, its minimum polynomial has degree at least 2. Then, since $\beta^2 - 16\beta + 4 = 0$ in K, the minimum polynomial of β is $X^2 - 16X + 4$.

Again, the minimum polynomials of $\alpha^3 - 14\alpha$ and $\alpha^2 = 18\alpha$ are at least 2. Note next that

$$(\alpha^3 - 14\alpha)^2 = \alpha^6 - 28\alpha^4 + 196\alpha^2$$

$$= \alpha^2(\alpha^4 - 16\alpha^2 + 4) - 12(\alpha^4 - 16\alpha^2 + 4) + 48 = 48 ,$$

and so the minimum polynomial of $\alpha^3 - 14\alpha$ is $X^2 - 48$. Similarly,

$$(\alpha^3 - 18\alpha)^2 = \alpha^6 - 36\alpha^4 + 324\alpha^2$$

$$= \alpha^2(\alpha^4 - 16\alpha^2 + 4) - 20(\alpha^4 - 16\alpha^2 + 4) + 80 = 80 ,$$

and so the minimum polynomial of $\alpha^3 - 18\alpha$ is $X^2 - 80$.

3.18. If $g = X^3 + X + 1$ were reducible, it would have a linear factor, and hence a root , either 0 or 1. Since neither 0 nor 1 is a root, g must be irreducible.

The elements of K are $0, 1, \alpha, 1 + \alpha, \alpha^2, 1 + \alpha^2, \alpha + \alpha^2, 1 + \alpha + \alpha^2$. Then $\alpha^3 = 1 + \alpha, \alpha^4 = \alpha + \alpha^2, \alpha^5 = 1 + \alpha + \alpha^2, \alpha^6 = 1 + \alpha^2, \alpha^7 = 1$, and so $K \setminus \{0\}$ is a cyclic group of order 7, generated by α. (It is indeed generated by any of its elements except 1.)

Chapter 4

4.1. Let $ABCD$ be a parallelogram.

Draw a circle with centre A passing through B, meeting the line AB again in P. Draw the perpendicular bisector of BP, meeting CD in Q. Similarly, draw the circle with centre B passing through A, meeting the line AB in R, and then draw the perpendicular bisector of AR, meeting CD in S. The rectangle $ABSQ$ has the same area as the parallelogram $ABCD$. Then construct a square equal in area to the rectangle $ABSQ$.

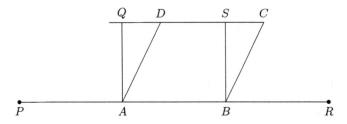

4.2. Let P, Q, R be non-collinear points, forming an angle QPR.

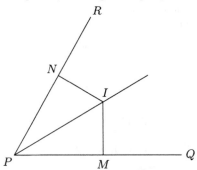

Draw a circle with centre P passing through Q, meeting PR in a point X (not shown). Then draw the perpendicular bisectors of PQ and PX. These meet in I, and PI is the required bisector.

4.3. Suppose that the angle is $\angle IOA$, where O is $(0,0)$, I is $(0,1)$ and A is (a,b) Let $K_0 = \mathbb{Q}(a,b)$. The circle with centre O passing though I meets OA in the point $C(a/\sqrt{a^2+b^2}, b/\sqrt{a^2+b^2})$. So we must extend K_0 to $K_1 = K_0(\sqrt{a^2+b^2})$.

The construction of the perpendicular bisector of OI involves the intersection of the circles $x^2 + y^2 = 1$ and $x^2 + y^2 = 2x$. The points of intersection are $(1/2, \pm\sqrt{3}/2)$, and so we must extend K_1 to $K_2 = K_1(\sqrt{3})$.

Similarly, the construction of the perpendicular bisector of OC involves the intersection of the circles $x^2 + y^2 = 1$ and $\left(x - (a/\sqrt{a^2+b^2})\right)^2 + \left(y - (b/\sqrt{a^2+b^2})\right)^2 = 1$. Subtracting the two equations and getting rid of fractions gives the equation $2ax + 2by = \sqrt{a^2+b^2}$ of the perpendicular bisector (the line joining the two points of intersection of the circles). After a bit of algebra, one finds that the two points of intersection are

$$\left(\frac{a \pm b\sqrt{3}}{\sqrt{a^2+b^2}}, \frac{b \mp a\sqrt{3}}{\sqrt{a^2+b^2}} \right).$$

The coordinates lie inside the field K_2. No further extensions are required when we find the coordinates $\left(1/2, [\sqrt{a^2 + b^2} - a]/(2b)\right)$ of the two perpendicular bisectors.

4.4. The intersection in the first quadrant of the circle with centre O passing through I and the circle with centre I passing through O is the point $P(1/2, \sqrt{3}/2)$, and the angle $\angle IOP$ is $\pi/3$. We can bisect this angle to obtain $\pi/6$.

As for $\pi/4$, from Example 4.4 we know how to construct the square $OIAB$, where $A = (1,1)$ and $B = (0,1)$. The angle IOA is $\pi/4$.

Chapter 5

5.1. First, $X^4 - 5X^2 + 6 = (X^2 - 2)(X^2 - 3)$, and the splitting field over $\mathbb{Q}$ is $\mathbb{Q}(\sqrt{2}, \sqrt{3})$. The degree of the extension is 4, and $\{1, \sqrt{2}, \sqrt{3}, \sqrt{6}\}$ is a basis.

Next, $X^4 - 1$ factorises over $\mathbb{C}$ into $(X + 1)(X - 1)(X + i)(X - i)$. The splitting field is $\mathbb{Q}(i)$, of degree 2 over $\mathbb{Q}$.

Finally, $X^4 + 1$ factorises over $\mathbb{C}$ into $(X - e^{i\pi/4})(X - e^{-i\pi/4})(X - e^{3i\pi/4})(X - e^{-3i\pi/4})$. In standard form the roots are $(1+i)/\sqrt{2}$, $(1-i)/\sqrt{2}$, $(-1+i)/\sqrt{2}$, $(-1-i)/\sqrt{2}$. The splitting field is $\mathbb{Q}(i, \sqrt{2})$, of degree 4 over $\mathbb{Q}$, with basis $\{1, i, \sqrt{2}, i\sqrt{2}\}$.

5.2. First, $X^6 - 1$ factorises over $\mathbb{C}$ into

$$(X - 1)(X + 1)(X - e^{i\pi/3})(X - e^{-i\pi/3})(X - e^{2i\pi/3})(X - e^{-2i\pi/3}).$$

In standard form the non-real roots are

$$(1 + i\sqrt{3})/2, \ (1 - i\sqrt{3})/2, \ (-1 - i\sqrt{3})/2, \ (-1 + i\sqrt{3})/2.$$

The splitting field is $\mathbb{Q}(i\sqrt{3})$, of degree 2 over $\mathbb{Q}$.

Next, $X^6 + 1$ factorises over $\mathbb{C}$ into

$$(X - i)(X + i)(X - e^{i\pi/6})(X - e^{-i\pi/6})(X - e^{5i\pi/6})(X - e^{-5i\pi/6}).$$

In standard form the roots are

$$i, \ -i, \ (\sqrt{3} + i)/2, \ (\sqrt{3} - i)/2, \ (-\sqrt{3} + i)/2, \ (-\sqrt{3} - i)/2.$$

The splitting field is $\mathbb{Q}(i, \sqrt{3})$, of degree 4 over $\mathbb{Q}$.

Finally, $X^6 - 27 = X^6 - (\sqrt{3})^6$, which factorises over $\mathbb{C}$ into

$$(X - \sqrt{3})(X + \sqrt{3})(X - \sqrt{3}e^{i\pi/3})(X - \sqrt{3}e^{-i\pi/3})$$
$$\times (X - \sqrt{3}e^{2i\pi/3})(X - \sqrt{3}e^{-2i\pi/3}).$$

In standard form the non-real roots are

$$(\sqrt{3} + 3i)/2, \ (\sqrt{3} - 3i)/2, \ (-\sqrt{3} - 3i)/2, \ (-sqrt3 - 3i)/2.$$

The splitting field is $\mathbb{Q}(i, \sqrt{3})$, of degree 4 over $\mathbb{Q}$.

5.3. Denote $\sqrt[4]{3}$ by α. Over $\mathbb{C}$, the polynomial factorises into

$$(X - \alpha e^{i\pi/4})(X - \alpha e^{-i\pi/4})(X - \alpha e^{3i\pi/4})(X - \alpha e^{-3i\pi/4}),$$

and, in standard form, the roots are

$$r_1 = \alpha(1 + i)/\sqrt{2}, \; r_2 = \alpha(1 - i)/\sqrt{2}, \; r_3 = \alpha(-1 + i)/\sqrt{2}, \; r_4 = \alpha(-1 - i)/\sqrt{2}.$$

The splitting field is generated over $\mathbb{Q}$ by these roots. It is clear that $r_1 = \frac{1}{2}(\alpha\sqrt{2} + i\alpha\sqrt{2}) \in \mathbb{Q}(i, \alpha\sqrt{2})$; similarly, $r_2.r_3, r_4 \in \mathbb{Q}(i, \alpha\sqrt{2})$. Thus $\mathbb{Q}(r_1, r_2, r_3, r_4) \subseteq \mathbb{Q}(i, \alpha\sqrt{2})$. Conversely, $r_1 + r_2 = \alpha\sqrt{2}$, and $(r_1 - r_2)/(r_1 + r_2) = \alpha i\sqrt{2}/\alpha\sqrt{2} = i$, and so $\mathbb{Q}(i, \alpha\sqrt{2}) \subseteq \mathbb{Q}(r_1, r_2, r_3, r_4)$. The minimum polynomial of $\alpha\sqrt{2}$ is $X^4 - 12$, and so $[\mathbb{Q}(\alpha\sqrt{2}) : \mathbb{Q}] = 4$. It follows that $[\mathbb{Q}(i, \alpha\sqrt{2}) : \mathbb{Q}] = 8$.

5.4. Since $f(0)$ and $f(1)$ are both non-zero, f has no linear factor. Hence f is irreducible. The field $K = \mathbb{Z}_2[X]/\langle f\rangle$ is generated over $\mathbb{Z}_2$ by $\alpha = X + \langle f\rangle$, and has eight elements

$$0, 1, \alpha, 1 + \alpha, \alpha^2, 1 + \alpha^2, \alpha + \alpha^2, 1 + \alpha + \alpha^2.$$

The multiplication table of the non-zero elements is

	1	α	$1+\alpha$	α^2	$1+\alpha^2$	$\alpha+\alpha^2$	$1+\alpha+\alpha^2$
1	1	α	$1+\alpha$	α^2	$1+\alpha^2$	$\alpha+\alpha^2$	$1+\alpha+\alpha^2$
α	α	α^2	$\alpha+\alpha^2$	$1+\alpha^2$	$1+\alpha+\alpha^2$	1	$1+\alpha$
$1+\alpha$	$1+\alpha$	$\alpha+\alpha^2$	$1+\alpha^2$	1	$1+\alpha+\alpha^2$	$1+\alpha+\alpha^2$	α^2
α^2	α^2	$1+\alpha^2$	1	$1+\alpha+\alpha^2$	$1+\alpha$	α	$\alpha+\alpha^2$
$1+\alpha^2$	$1+\alpha^2$	$1+\alpha+\alpha^2$	α	$1+\alpha$	$\alpha+\alpha^2$	α^2	1
$\alpha+\alpha^2$	$\alpha+\alpha^2$	1	$1+\alpha+\alpha^2$	α	α^2	$1+\alpha$	$1+\alpha^2$
$1+\alpha+\alpha^2$	$1+\alpha+\alpha^2$	$1+\alpha$	$1+\alpha$	α^2	$\alpha+\alpha^2$	$1+\alpha^2$	α

By trial and error, comparing ζ^3 with $1 + \zeta^2$ for each of the seven elements, we find the three roots of f in K. Thus f splits completely in $K[X]$:

$$f = (X + \alpha)(X + \alpha^2)(X + 1 + \alpha + \alpha^2).$$

Chapter 6

6.1. It is easy to verify the identity for small values of $m + n$. Suppose that $D(fg) = (Df)g + f(Dg)$ for all polynomials such that $\partial f + \partial g < k$. Let $f = a_0 + a_1 X + \cdots + a_m X^m$, where $m > 1$, let $g = b_0 + b_1 X + \cdots b_n X^n$, and let $m + n = k$. Write $f = f_1 + f_2$, where $f_2 = a_m X^m$. Then $D(fg) = D(f_1 g + f_2 g) = D(f_1 g) + D(f_2 g)$. Now $D(f_1 g) = (Df_1)g + f_1(Dg)$ by the induction hypothesis. Also,

$$\begin{aligned}
D(f_2 g) &= D(a_m b_0 X^m + a_m b_1 X^{m+1} + \cdots a_m b_n X^{m+n}) \\
&= a_m(m b_0 X^{m-1} + (m+1)b_1 X^m + (m+2)b_2 X^{m+1} + \\
&\quad + \cdots + (m+n)b_n X^{m+n-1}) \\
&= m a_m X^{m-1}(b_0 + b_1 X + \cdots + b_n X^n) + \\
&\quad + a_m X^m(b_1 + 2b_2 X + \cdots + n b_n X^{n-1}) \\
&= (Df_2)g + f_2(Dg).
\end{aligned}$$

Hence

$$D(fg) = D(f_1g) + D(f_2g) = (Df_1)g + f_1(Dg) + (Df_2)g + f_2(Dg)$$
$$= \Big(D(f_1 + f_2)\Big)g + (f_1 + f_2)(Dg) = (Df)g + f(Dg).$$

6.2. The result is certainly true for $n = 1$. Then, by the product rule,

$$D[(X - \alpha)^n] = (X - \alpha)D[(X - \alpha)^{n-1}] + (X - \alpha)^{n-1}$$
$$= (X - \alpha)(n - 1)(X - \alpha)^{n-2} + (X - \alpha)^{n-1}$$
$$= n(X - \alpha)^{n-1}.$$

6.3. There are p^2 quadratic polynomials in $\mathbb{Z}_p[X]$. There are p linear polynomials. Multiplying the linear polynomials together involves p squares, and $\binom{p}{2} = p(p-1)/2$ products of distinct factors. Hence the number of irreducible polynomials is $p^2 - p - p(p-1)/2 = p(p-1)/2$.

6.4. If $X^2 + 2$ were reducible, it would have a linear factor, and hence a root in $\mathbb{Z}_5$. Checking each of $0, \pm 1, \pm 2$, we find that there is no such root. The element α has the property that $\alpha^2 = -2$, and the order of $1 + \alpha$ is a divisor of 24. Thus $o(1 + \alpha) \in \{2, 4, 8, 3, 6, 12, 24\}$. Now

$$(1 + \alpha)^2 = 1 + 2\alpha - 2 = -1 + 2\alpha,$$
$$(1 + \alpha)^4 = (-1 + 2\alpha)^2 = 1 - 4\alpha + 4\alpha^2 = 1 + \alpha - \alpha^2 = -2 + \alpha,$$
$$(1 + \alpha)^8 = (-2 + \alpha)^2 = 4 - 4\alpha + \alpha^2 = -1 + \alpha - 2 = 2 + \alpha,$$

and so $o(1 + \alpha) \notin \{2, 4, 8\}$ Also

$$(1 + \alpha)^3 = (-1 + 2\alpha)(1 + \alpha) = -1 + \alpha + 2\alpha^2 = -1 + \alpha - 4 = \alpha,$$
$$(1 + \alpha)^6 = \alpha^2 = -2,$$
$$(1 + \alpha)^{12} = 4 = -1,$$

and so $o(1 + \alpha) \notin \{3, 6, 12\}$. From the last of the above equations we see that $o(1 + \alpha) = 24$.

6.5. It is easy to see that $f = X^4 + X + 1$ has no linear factor, since $f(0) = f(1) = 1$. The quadratic polynomials in $\mathbb{Z}_2[X]$ are X^2, $X^2 + 1$, $X^2 + X$ and $X^2 + X + 1$, and of these only $X^2 + X + 1$ is irreducible. So if f were to have quadratic factors it could only be $(X^2 + X + 1)^2 = X^4 + X^2 + 1$. Hence $X^4 + X + 1$ must be irreducible.

Since $\alpha^4 = \alpha + 1$, the positive powers of α are

n	2	3	4	5	6	7	8
α^n	α^2	α^3	$1 + \alpha$	$\alpha + \alpha^2$	$\alpha^2 + \alpha^3$	$1 + \alpha + \alpha^3$	$1 + \alpha^2$

n	9	10	11	12
α^n	$\alpha + \alpha^3$	$1 + \alpha + \alpha^2$	$\alpha + \alpha^2 + \alpha^3$	$1 + \alpha^2 + \alpha^3$

n	13	14	15
α^n	$1 + \alpha + \alpha^2 + \alpha^3$	$1 + \alpha^3$	1

6.6. (i) Since $\varphi(ab) = (ab)^p = a^p b^b = \varphi(a)\varphi(b)$ and $\varphi(a+b) = (a+b)^p = a^p + b^p = \varphi(a) + \varphi(b)$, the map is a homomorphism. Also, since $\varphi(a) = \varphi(b) \Rightarrow 0 = \varphi(a) - \varphi(b) = a^p - b^p = (a-b)^p$, we deduce that $a - b = 0$. Thus φ is a monomorphism. Thus $|\varphi(F)| = |F|$, and this implies $\varphi(F) = F$ if F is finite. Hence φ is an automorphism in this case.

The elements of $\mathbb{Z}_p$ are $0, 1, 1+1, 1+1+1, \ldots$ Certainly $\varphi(0) = 0$ and $\varphi(1) = 1$. Then, for example, $\varphi(1+1+1) = \varphi(1) + \varphi(1) + \varphi(1) = 1+1+1$. So φ is the identity map.

(ii) Consider the field $\mathbb{Z}_p(X)$ of all rational forms over $\mathbb{Z}_p$. Let f be a non-constant monic polynomial. Then $\varphi(f) = f^p$ is of degree $p\partial f$, and so, for example, no polynomial of degree 1 is in the image of φ.

6.7. (i) $\varphi(\alpha) = \alpha^5 = -\alpha$.
(ii) $\varphi(\alpha) = \alpha^2$, $\varphi^2(\alpha) = 1 + \alpha$, $\varphi^3(\alpha) = 1 + \alpha^2$. (The map $\varphi^4 : x \mapsto x^{16}$ is the identity map.)

Chapter 7

7.1. (i) Since $0 + 0 = 0$, we have that $\alpha(0) = 0 + \alpha(0) = [-(\alpha(0)) + \alpha(0)] + \alpha(0) = -(\alpha(0)) + [\alpha(0) + \alpha(0)] = -(\alpha(0)) + \alpha(0 + 0) = -(\alpha(0)) + \alpha(0) = 0$.

$\alpha(-x) = 0 + \alpha(-x) = [-(\alpha(x)) + \alpha(x)] + \alpha(-x) = -(\alpha(x)) + [\alpha(x) + \alpha(-x)] = -(\alpha(x)) + \alpha(x + (-x)) = -(\alpha(x)) + \alpha(0) = -(\alpha(x)) + 0 = -(\alpha(x))$.

(ii) The multiplicative statements follow by similar arguments.

7.2. Let $\varphi \in \operatorname{Aut} Q$. Then $\varphi(1) = 1$ and, by the previous exercise, $\varphi(-1) = -1$. It follows that, for all n in $\mathbb{N}$,

$$\varphi(n) = \varphi(1 + 1 + \cdots + 1) = 1 + 1 + \cdots + 1 = n$$

and, similarly, that $\varphi(-n) = -n$. If $m, n \in \mathbb{Z}$ and $n \neq 0$, then, by the previous exercise,

$$\varphi(\tfrac{m}{n}) = \varphi(m)[\varphi(n)]^{-1} = \tfrac{m}{n}.$$

So $\operatorname{Aut}(\mathbb{Q})$ is the trivial group. By a simpler version of the above argument, we have same result for $\mathbb{Z}_p$.

7.3. Let E be a subfield of L containing K and let H be a subgroup of $\operatorname{Gal}(L : K)$. From Theorem 7.6 we know that $E \subseteq \varPhi(\Gamma(E))$, and by the order-reversing property it then follows that $\Gamma(E) \supseteq (\Gamma\varPhi\Gamma)(E)$. On the other hand, we know that $H \subseteq \Gamma(\varPhi(H))$; and so, substituting $\Gamma(E)$ for H, we see that $\Gamma(E) \subseteq (\Gamma\varPhi\Gamma)(E)$. Hence $\Gamma\varPhi\Gamma = \Gamma$.

Similarly, from $H \subseteq \Gamma(\varPhi(H))$ we have, by the order-reversing property, that $\varPhi(H) \supseteq (\varPhi\Gamma\varPhi)(H)$. On the other hand, substituting $\varPhi(H)$ for E in $E \subseteq \varPhi(\Gamma(E))$ gives $\varPhi(H) \subseteq (\varPhi\Gamma\varPhi)(H)$. Hence $\varPhi\Gamma\varPhi = \varPhi$.

7.4. The map τ is given by

$$\tau(a + b\sqrt{2} + ci\sqrt{3} + di\sqrt{6}) = a - b\sqrt{2} + ci\sqrt{3} - di\sqrt{6}.$$

It is clear that τ is its own inverse, and so τ is one-one and onto. Let

$$z_j = a_j + b_j\sqrt{2} + c_j i\sqrt{3} + d_j i\sqrt{6} \quad (j = 1, 2).$$

The proof that $\tau(z_1 + z_2) = \tau(z_1) + \tau(z_2)$ is routine. Also,

$$z_1 z_2 = (a_1 a_2 + 2b_1 b_2 - 3c_1 c_2 - 6d_1 d_2) + (a_1 b_2 + a_2 b_1 - 3c_1 d_2 - 3c_2 d_1)\sqrt{2}$$
$$+ (a_1 c_2 + a_2 c_1 + 2b_1 d_2 + 2b_2 d_1)i\sqrt{3} + (a_1 d_2 + a_2 d_1 + b_1 c_2 + b_2 c_1)i\sqrt{6}.$$

By a similar calculation, we find that $\tau(z_1)\tau(z_2)$ is equal to

$$(a_1 a_2 + 2b_1 b_2 - 3c_1 c_2 - 6d_1 d_2) - (a_1 b_2 + a_2 b_1 - 3c_1 d_2 - 3c_2 d_1)\sqrt{2}$$
$$+ (a_1 c_2 + a_2 c_1 + 2b_1 d_2 + 2b_2 d_1)i\sqrt{3} - (a_1 d_2 + a_2 d_1 + b_1 c_2 + b_2 c_1)i\sqrt{6},$$

and this coincides with $\tau(z_1 z_2)$. Thus τ is an automorphism.

7.5.   It is clear that $K = \mathbb{Q}(i+\sqrt{2}) \subseteq \mathbb{Q}(i, \sqrt{2})$. In fact the two fields are identical, for $(i+\sqrt{2})^2 = 1 + 2i\sqrt{2}$, $(i+\sqrt{2})^3 = 5i - \sqrt{2}$, and so $i = \frac{1}{6}[(i+\sqrt{2})^3 + (i+\sqrt{2})] \in K$ and $\sqrt{2} = \frac{1}{6}[5(i + \sqrt{2}) - (i + \sqrt{2})^3] \in K$.

Any $\mathbb{Q}$-automorphism of $K = \mathbb{Q}(i, \sqrt{2})$ must map i to $\pm i$ and $\sqrt{2}$ to $\pm\sqrt{2}$. So there are 4 elements of $\mathrm{Gal}(K, \mathbb{Q})$, given by

$$\iota : i \mapsto i,\ \sqrt{2} \mapsto \sqrt{2} \qquad \varphi : i \mapsto -i,\ \sqrt{2} \mapsto \sqrt{2}$$
$$\psi : i \mapsto i,\ \sqrt{2} \mapsto -\sqrt{2} \qquad \chi : i \mapsto -i,\ \sqrt{2} \mapsto -\sqrt{2}.$$

The multiplication is given in the Cayley table:

	ι	φ	ψ	χ
ι	ι	φ	ψ	χ
φ	φ	ι	χ	ψ
ψ	ψ	χ	ι	φ
χ	χ	ψ	φ	ι

7.6.   GF(8) is $\mathbb{Z}_2[X]/\langle X^3 + X + 1\rangle$. If $\alpha = X + \langle X^3 + X + 1\rangle$, then we may write GF(8) as $\mathbb{Z}_2(\alpha)$, and the elements of GF(8) are $0, 1, \alpha, 1+\alpha, \alpha^2, 1+\alpha^2, \alpha+\alpha^2, 1+\alpha+\alpha^2$. The powers of α are given by

n	1	2	3	4	5	6	7
α^n	α	α^2	$1+\alpha$	$\alpha+\alpha^2$	$1+\alpha+\alpha^2$	$1+\alpha^2$	1

Since $\alpha^3 + \alpha + 1 = 0$, it follows, by squaring, that $\alpha^6 + \alpha^2 + 1 = 0$, and so α^2 is also a root of $X^3 + X + 1$. Squaring again, we see that $\alpha^4 = \alpha + \alpha^2$ is again a root of $X^3 + X + 1$. Any $\mathbb{Z}_2$-automorphism must map a root of $X^3 + X + 1$ to another root. Accordingly, there are three elements in $\mathrm{Gal}(\mathrm{GF}(8), \mathbb{Z}_2)$:

$$\iota : \alpha \mapsto \alpha, \quad \varphi : \alpha \mapsto \alpha^2, \quad \psi : \alpha \mapsto \alpha + \alpha^2,$$

and the multiplication table is

	ι	φ	ψ
ι	ι	φ	ψ
φ	φ	ψ	ι
ψ	ψ	ι	φ

7.7.   Since L is a normal extension of K, it is a splitting field for some polynomial f in $K[X]$. Since $f \in E[X]$, we conclude that L is a normal extension of E.

7.8. The minimum polynomial of $u = \sqrt[4]{2}$ is $X^4 - 2$ (which is irreducible over $\mathbb{Q}$ by the Eisenstein criterion). Over $\mathbb{C}$, the polynomial factorises as $(X - u)(X + u)(X - iu)(X + iu)$, and so $\mathbb{Q}(u, i)$ is a normal extension of $\mathbb{Q}(u)$. Over any field K such that $\mathbb{Q}(u) \subseteq K \subset \mathbb{Q}(u, i)$, the polynomial $X^4 - 2$ has a root but does not split completely; hence K is not normal. It follows that $\mathbb{Q}(u, i)$ is the normal closure.

7.9. Let $u = f/g$, $v = p/q$. Then $u + v = (fq + gp)/(gq)$, $uv = (fp)/(gq)$, $u/v = (fq)/(gp)$.

$$D(u + v) = \frac{1}{(gq)^2}[gqD(fq + gp) - (fq + gp)D(gq)]$$

$$= \frac{1}{(gq)^2}[gqfDq + gq^2Df + g^2qDp + gqpDg$$

$$- fqgDq - fq^2Dg - g^2pDq - gpqDg]$$

$$= \frac{1}{(gq)^2}[gq^2Df + g^2qDp - fq^2Dg - g^2pDq]$$

$$= \frac{1}{g^2q^2}[q^2(gDf - fDg) + g^2(qDp - pDq)]$$

$$= \frac{gDf - fDg}{g^2} + \frac{qDp - pDq}{q^2}$$

$$= Du + Dv,$$

$$D(uv) = \frac{1}{(gq)^2}[gqD(fp) - fpD(gq)]$$

$$= \frac{1}{(gq)^2}[gqfDp + gqpDf - fpgDq - fpqDg]$$

$$vDu + uDv = \frac{p(gDf - fDg)}{g^2q} + \frac{f(qDp - pDq)}{gq^2}$$

$$= \frac{1}{(gq)^2}[pgqDf - pfqDg + gfqDp - gfpDq]$$

$$= D(uv).$$

Next,

$$D(1/v) = D(q/p) = \frac{pDq - qDp}{p^2} = -\frac{q^2}{p^2}\frac{qDp - pDq}{q^2} = -\frac{1}{v^2}Dv,$$

and so, by the product rule

$$D(u/v) = uD(1/v) + (1/v)Du = -\frac{uDv}{v^2} + \frac{Du}{v} = \frac{vDu - uDv}{v^2}.$$

7.10. Suppose that φ is an automorphism. The only candidates for an irreducible inseparable polynomial are polynomials of the type $f = a_0 + a_1X^p + \cdots + a_nX^{np}$, with $a_0, a_1, \ldots, a_n \in F$. By our assumption, $a_i = b_i^p$ for some b_i ($i = 0, 1, \ldots, n$). Thus $f = (b_0 + b_1X + \cdots + b_nX^n)^p$ is not irreducible.

Conversely, suppose that φ is not an automorphism, and let a be an element not in the image of φ. By Theorem 7.24, the polynomial $X^p - a$ is irreducible and inseparable, and so F is not perfect.

7.11. Let $z \in K$, with minimum polynomial f of degree k. Since z is inseparable, we may suppose that we have an irreducible polynomial $f = a_{mp}X^{mp} + \cdots + a_p X^p + a_0$ such that $f(z) = 0$. Hence, by Theorem 7.24, we have an irreducible polynomial $f_1 = a_{mp}X^m + \cdots + a_p X + a_0$ such that $f_1(z^p) = 0$. But z^p is also inseparable, and so in fact

$$f_1 = a_{rp^2}X^{rp} + \cdots + a_{p^2}X^p \,.$$

We can continue this process, reaching the conclusion that the minimum polynomial of $z^{p^{s-1}}$ contains only powers X^i for which $p^s \mid i$. We continue until only one non-constant term is left, which happens when $p^n \leq k < p^{n+1}$, and obtain a minimum polynomial $X^{p^n} + a$ for z.

7.12. Let $\alpha \in G = \mathrm{Gal}(\mathbb{Q}(u, i\sqrt{3}) : \mathbb{Q})$. By Theorem 7.9, $\alpha(i\sqrt{3}) = \pm i\sqrt{3}$ and $\alpha(u) \in \{u, ue^{2i\pi/3}, ue^{-2i\pi/3}\}$. Since every automorphism is determined by its effect on u and $i\sqrt{3}$, there are precisely 6 automorphisms in the Galois group, namely,

$$
\begin{array}{lll}
\iota \; : \; u \mapsto u & \alpha \; : \; u \mapsto ue^{2i\pi/3} & \beta \; : \; u \mapsto ue^{-2i\pi/3} \\
 \; i\sqrt{3} \mapsto i\sqrt{3} & \; i\sqrt{3} \mapsto i\sqrt{3} & \; i\sqrt{3} \mapsto i\sqrt{3} \\
\\
\lambda \; : \; u \mapsto u & \mu \; : \; u \mapsto ue^{2i\pi/3} & \nu \; : \; \mapsto ue^{-2i\pi/3} \\
 \; i\sqrt{3} \mapsto -i\sqrt{3} & \; i\sqrt{3} \mapsto -i\sqrt{3} & \; i\sqrt{3} \mapsto -i\sqrt{3} \,.
\end{array}
$$

The multiplication in the group is given by the table

	ι	α	β	λ	μ	ν
ι	ι	α	β	λ	μ	ν
α	α	β	ι	μ	ν	λ
β	β	ι	α	ν	λ	μ
λ	λ	ν	μ	ι	β	α
μ	μ	λ	ν	α	ι	β
ν	ν	μ	λ	β	α	ι

This takes a bit of computation: for example, $(\alpha\lambda)(u) = \alpha(u) = ue^{2i\pi/3} = \mu(u)$ and $(\alpha\lambda)(i\sqrt{3}) = \alpha(-i\sqrt{3}) = -i\sqrt{3} = \mu(i\sqrt{3})$, while $(\lambda\alpha)(u) = \lambda(ue^{2i\pi/3}) = \lambda(u)\lambda(e^{2i\pi/3}) = ue^{-2i\pi/3} = \nu(u)$ (since $\lambda(e^{2i\pi/3}) = \lambda((1 + i\sqrt{3})/2) = (1 - i\sqrt{3})/2 = e^{-2i\pi/3}$) and $(\lambda\alpha)(i\sqrt{3}) = \lambda(i\sqrt{3}) = -i\sqrt{3} = \nu(i\sqrt{3})$.
The proper subgroups are $H_1 = \{\iota, \alpha, \beta\}$, $H_2 = \{\iota, \lambda\}$, $H_3 = \{\iota, \mu\}$ and $H_4 = \{\iota, \nu\}$; and $\Phi(H_1) = \mathbb{Q}(i\sqrt{3})$, $\Phi(H_2) = \mathbb{Q}(u)$, $\Phi(H_3) = \mathbb{Q}(ue^{-2i\pi/3})$, $\Phi(H_4) = \mathbb{Q}(ue^{2i\pi/3})$.

7.13. The group G has 8 elements:

$$
\begin{array}{llll}
\iota : \sqrt{2} \mapsto \sqrt{2}, & \sqrt{3} \mapsto \sqrt{3}, & \sqrt{5} \mapsto \sqrt{5} \\
\alpha : \sqrt{2} \mapsto -\sqrt{2}, & \sqrt{3} \mapsto \sqrt{3}, & \sqrt{5} \mapsto \sqrt{5} \\
\beta : \sqrt{2} \mapsto \sqrt{2}, & \sqrt{3} \mapsto -\sqrt{3}, & \sqrt{5} \mapsto \sqrt{5} \\
\gamma : \sqrt{2} \mapsto \sqrt{2}, & \sqrt{3} \mapsto \sqrt{3}, & \sqrt{5} \mapsto -\sqrt{5} \\
\lambda : \sqrt{2} \mapsto \sqrt{2}, & \sqrt{3} \mapsto -\sqrt{3}, & -\sqrt{5} \mapsto \sqrt{5} \\
\mu : \sqrt{2} \mapsto -\sqrt{2}, & \sqrt{3} \mapsto \sqrt{3}, & \sqrt{5} \mapsto -\sqrt{5} \\
\nu : \sqrt{2} \mapsto -\sqrt{2}, & \sqrt{3} \mapsto -\sqrt{3}, & \sqrt{5} \mapsto \sqrt{5} \\
\rho : \sqrt{2} \mapsto -\sqrt{2}, & \sqrt{3} \mapsto -\sqrt{3}, & \sqrt{5} \mapsto -\sqrt{5} \,.
\end{array}
$$

The multiplication table is

	ι	α	β	γ	λ	μ	ν	ρ
ι	ι	α	β	γ	λ	μ	ν	ρ
α	α	ι	ν	μ	ρ	γ	β	λ
β	β	ν	ι	λ	γ	ρ	α	μ
γ	γ	μ	λ	ι	β	α	ρ	ν
λ	λ	ρ	γ	β	ι	ν	μ	α
μ	μ	γ	ρ	α	ν	ι	λ	β
ν	ν	β	α	ρ	μ	λ	ι	γ
ρ	ρ	λ	μ	ν	α	β	γ	ι

The subgroups of order 4, with their images under Φ, are

$$H_1 = \{\iota, \beta, \gamma, \lambda\}, \ \Phi(H_1) = \mathbb{Q}(\sqrt{2}); \ H_2 = \{\iota, \alpha, \gamma, \mu\}, \ \Phi(H_2) = \mathbb{Q}(\sqrt{3});$$

$$H_3 = \{\iota, \alpha, \beta, \nu\}, \ \Phi(H_3) = \mathbb{Q}(\sqrt{5}); \ H_4 = \{\iota, \nu, \gamma, \rho\}, \ \Phi(H_4) = \mathbb{Q}(\sqrt{6});$$

$$H_5 = \{\iota, \mu, \beta, \rho\}, \ \Phi(H_5) = \mathbb{Q}(\sqrt{10}); \ H_6 = \{\iota, \lambda, \alpha, \rho\}, \ \Phi(H_6) = \mathbb{Q}(\sqrt{15});$$

$$H_7 = \{\iota, \lambda, \mu, \nu\}, \ \Phi(H_7) = \mathbb{Q}(\sqrt{30}).$$

The subgroups of order 2, with their images under Φ, are

$$K_1 = \{\iota, \alpha\}, \ \Phi(K_1) = \mathbb{Q}(\sqrt{3}, \sqrt{5}); \ K_2 = \{\iota, \beta\}, \ \Phi(K_2) = \mathbb{Q}(\sqrt{2}, \sqrt{5});$$

$$K_3 = \{\iota, \gamma\}, \ \Phi(K_3) = \mathbb{Q}(\sqrt{2}, \sqrt{3}); \ K_4 = \{\iota, \lambda\}, \ \Phi(K_4) = \mathbb{Q}(\sqrt{2}, \sqrt{15});$$

$$K_5 = \{\iota, \mu\}, \ \Phi(K_5) = \mathbb{Q}(\sqrt{3}, \sqrt{10}); \ K_6 = \{\iota, \nu\}, \ \Phi(K_2) = \mathbb{Q}(\sqrt{5}, \sqrt{6});$$

$$K_7 = \{\iota, \rho\}, \ \Phi(K_7) = \mathbb{Q}(\sqrt{6}, \sqrt{10}).$$

Chapter 8

8.1. Here $a = 1$ and $b = -3$, and so $\Delta = 13$. Hence $q^3 = \frac{1}{2}(3 + \sqrt{13})$ and $r^3 = \frac{1}{2}(3 - \sqrt{13})$. If we take q and r as the real cube roots of q^3 and r^3, respectively, then $qr = -1$, as required. So the roots are $q + r$, $q\omega + r\omega^2$ and $q\omega^2 + r\omega$, where

$$q = [\frac{1}{2}(3 + \sqrt{13})]^{1/3}, \quad r = [\frac{1}{2}(3 - \sqrt{13})]^{1/3}.$$

8.2. Here $a = -1$ and $b = 1$, and so $\Delta = -3$. Hence $q^3 = \frac{1}{2}(-1 + i\sqrt{3}) = e^{2\pi i/3}$ and $r^3 = \frac{1}{2}(-1 - i\sqrt{3}) = e^{-2\pi i/3}$. We take $q = e^{2\pi i/9}$ and $r = e^{-2\pi i/9}$ and obtain the root $q + r = 2\cos(2\pi/9)$. The other roots are $q\omega + r\omega^2 = e^{8\pi i/9} + e^{-8\pi i/9} = 2\cos(8\pi/9)$, and $q\omega^2 + r\omega = 2\cos(4\pi/9)$. Notice that the roots are all real, but that we have had to use complex numbers to find them.

8.3. $X^{2p} - 1 = (X^p - 1)(X^p + 1) = (X - 1)(X + 1)(X^{p-1} + X^{p-2} + \cdots + X + 1)(X^{p-1} - X^{p-2} + \cdots - X + 1)$. Since the first three factors are (respectively) Φ_1, Φ_2 and Φ_p, the remaining factor must be Φ_{2p}.

8.4. $X^{15} - 1$ has factors $\Phi_1 = X - 1$, $\Phi_3 = X^2 + X + 1$ and $\Phi_5 = X^4 + X^3 + X^2 + X + 1$, and so $X^{15} - 1 = (X - 1)(X^2 + X + 1)(X^4 + X^3 + X^2 + X + 1)\Phi_{15}$. Note also that $X^{15} - 1 = (X^5)^3 - 1 = (X^5 - 1)(X^{10} + X^5 + 1) = (X - 1)(X^4 + X^3 + X^2 + X + 1)(X^{10} + X^5 + 1)$. Comparing the two factorisations, we deduce that $\Phi_{15} = (X^{10} + X^5 + 1)/(X^2 + X + 1)$, which equals (after a tedious calculation) $X^8 - X^7 + X^5 - X^4 + X^3 - X + 1$.

8.5.   It is clear from the definition that cyclotomic polynomials are monic. Suppose that
$$X^m - 1 = (a_0 + a_1 X + \cdots + a_p X^p)(b_0 + b_1 X + \cdots + b_q X^q)\,,$$
where $p + q = m$, $a_p = 1$ and $a_1, \ldots, a_{p-1} \in \mathbb{Z}$. Then, equating coefficients of X^m, we see that $1 = a_p b_q = b_q$, and so certainly $b_q \in \mathbb{Z}$. Suppose inductively that $b_{r+1}, \ldots, b_q \in \mathbb{Z}$. Then, equating coefficients of X^{p+r} gives
$$0 = a_p b_r + a_{p-1} b_{r+1} + \cdots + a_{p-q+r} b_q\,,$$
where $a_i = 0$ if $i < 0$. Thus
$$b_r = a_p b_r = -(a_{p-1} b_{r+1} + \cdots + a_{p-q+r} b_q) \in \mathbb{Z}\,.$$
Hence $b_j \in \mathbb{Z}$ for all j.

If we assume inductively that $\Phi_d \in \mathbb{Z}[X]$ for all $d < m$, and denote the set of divisors of m by Δ_m, we deduce from
$$X^m - 1 = \left(\prod_{d \in \Delta_m \setminus \{m\}} \Phi_d \right) \Phi_m$$
that $\Phi_m \in \mathbb{Z}[X]$.

8.6.   (i)   Let K be the splitting field in $\mathbb{C}$ of $X^{12} - 1$. It contains $\omega = e^{\pi i}/6$, and the Galois group has four elements, defined by
$$\omega \mapsto \omega, \ \omega \mapsto \omega^5 ; \omega \mapsto \omega^7 ; \omega \mapsto \omega^{11}\,.$$
It is isomorphic to the multiplicative group $\{\bar{1}, \bar{5}, \bar{7}, \bar{11}\}$ mod 12.

   (ii)   In the same way, the Galois group of $X^{15} - 1$ is isomorphic to the multiplicative group $\{\bar{1}, \bar{2}, \bar{4}, \bar{7}, \bar{8}, \bar{11}, \bar{13}, \bar{14}\}$ mod 15.

8.7.   (i)   If $x = z - \tau(z)$, then $\mathrm{Tr}_{K/F}(x) = (z - \tau(z)) + (\tau(z) - \tau^2(z)) + \cdots + (\tau^{n-1}(z) - \tau^n(z)) = z - \tau^n(z) = 0$. Conversely, suppose that $\mathrm{Tr}_{K/F}(x) = 0$. Then
$$-x = \tau(x) + \tau^2(x) + \cdots + \tau^{n-1}(x)\,.$$
As in the proof of Theorem 8.17, there exists t in K such that
$$u = x\tau(t) + (x + \tau(x))\tau^2(t) + \cdots$$
$$+ (x + \tau(x) + \tau^2(x) + \cdots + \tau^{n-2}(x))\tau^{n-1}(t)$$
is non-zero. Hence
$$\tau(u) = \tau(x)\tau^2(t) + (\tau(x) + \tau^2(x))\tau^3(t) + \cdots$$
$$\cdots + (\tau(x) + \tau^2(x) + \tau^3(x) + \cdots + \tau^{n-1}(x))\tau^n(t)$$
$$= \tau(x)\tau^2(t) + (\tau(x) + \tau^2(x))\tau^3(t) + \cdots + (-xt)\,,$$

and

$$u - \tau(u) = xt + x\tau(t) + (x + \tau(x))\tau^2(t) + (x + \tau(x) + \tau^2(x))\tau^3(t)$$
$$\cdots + (x + \tau(x) + \tau^2(x) + \cdots + \tau^{n-2}(x))\tau^{n-1}(t)$$
$$- \tau(x)\tau^2(t) - (\tau(x) + \tau^2(x))\tau^3(t) - \cdots$$
$$\cdots - (\tau(x) + \tau^2(x) + \tau^3(x) + \cdots + \tau^{n-2}(x))\tau^{n-1}(t)$$
$$= x(t + \tau(t) + \tau^2(t) + \cdots + \tau^{n-1}(t)) = x\mathrm{Tr}_{K/F}(t).$$

Since $\mathrm{Tr}_{K/F}(t) \in F$, by Theorem 8.16, it is left fixed by τ. Let $z = u/\mathrm{Tr}_{K/F}(t)$; then $z - \tau(z) = (u - \tau(u))/\mathrm{Tr}_{K/F}(t) = x$.

(ii) $z - \tau(z) = z' - \tau(z') \iff \tau(z - z') = z - z' \iff z - z' \in F$.

8.8. Let r be a root of $X^p - a$ in a splitting field K. Then the roots of $X^p - a$ in K are

$$r, r\omega, \ldots, r\omega^{p-1},$$

where ω is a primitive pth root of unity. A typical element of the Galois group $\mathrm{Gal}(K, F)$ is $\sigma_{s,t}$, where $\sigma_{s,t}(r) = r\omega^s$, $\sigma_{s,t}(\omega) = \omega^t$ (where $s = 0, 1, \ldots, p-1$ and $t = 1, 2, \ldots, p-1$), and, as in Example 8.22, $\sigma_{s,t}\sigma_{u,v} = \sigma_{s+tu,tv}$. If $\beta = \sigma_{1,1}$ and $\gamma = \sigma_{0,w}$, where w is an element of order $p-1$ in the (cyclic) multiplicative group of non-zero integers mod p, then $\beta^p = \gamma^{p-1} = 1$, and $\gamma\beta = \sigma_{w,w} = \beta^w\gamma$. The group, of order $p(p-1)$ has presentation $\langle \beta, \gamma \mid \beta^p = \gamma^{p-1} = \beta^w\gamma\beta^{-1}\gamma^{-1} = 1 \rangle$.

8.9. The 6th roots of unity are 1, -1, $e^{\pm\pi i/3} = \frac{1}{2}(1 \pm i\sqrt{3})$, $e^{\pm 2\pi i/3} = \frac{1}{2}(-1 \pm i\sqrt{3})$, and so (writing $e^{\pi i/3}$ as ω) we deduce that $\mathbb{Q}(\omega) = \mathbb{Q}(i\sqrt{3})$. The primitive roots of the equation are ω and $\omega^5 = \bar{\omega}$. It is clear that, over $\mathbb{Q}(i\sqrt{3})$, the polynomial $X^6 + 3$ splits completely as $(X^3 + i\sqrt{3})(X^3 - i\sqrt{3})$. For suppose that $X^3 - i\sqrt{3}$ is not irreducible over $\mathbb{Q}(i\sqrt{3})$. Then it has a linear factor, and so there exists $a + bi\sqrt{3}$, with $a, b \in \mathbb{Q}$, such that $i\sqrt{3} = (a + bi\sqrt{3})^3 = a^3 + 3a^2bi\sqrt{3} - 9ab^2 - 3b^3i\sqrt{3}$. Hence $a^3 - 9ab^2 = 0$ and $3a^2b - 3b^3 = 1$. If $a = 0$, then $-3b^3 = 1$, which is not possible for a rational b. Otherwise $a^2 - 9b^2 = 0$ and so $a = \pm 3b$. Hence $27b^3 - 3b^3 = 1$, and again this is not possible for a rational b.
The roots of $X^3 - i\sqrt{3}$ are $r, r\omega^2, r\omega^4$. The Galois group consists of elements $\sigma_{s,t}$, where $s \in \{0, 2, 4\}$ and $t \in \{1, -1\}$, defined by

$$\sigma_{s,t}(r) = r\omega^s, \quad \sigma_{s,t}(\omega) = \omega^t.$$

Then $\sigma_{s,t}\sigma_{u,v}(r) = \sigma_{s,t}(r\omega^u) = r\omega^s\omega^{tu} = r\omega^{s+tu}$, and $\sigma_{s,t}\sigma_{u,v}(\omega) = \sigma_{s,t}(\omega^v) = \omega^{tv}$, and so (mod 6) $\sigma_{s,t}\sigma_{u,v} = \sigma_{s+tu,tv}$. Note that $(\sigma_{2,1})^2 = \sigma_{4,1}$, $(\sigma_{2,1})^3 = 1$, and that $(\sigma_{0,-1})^2 = 1$. Notice also that $\sigma_{2,1}\sigma_{0,-1} = \sigma_{2,-1}$ and $\sigma_{0,-1}\sigma_{2,1} = \sigma_{4,-1} = \sigma_{4,1}\sigma_{0,-1}$. Writing $\sigma_{2,1}$ as β and $\sigma_{0,-1}$ as α gives

$$\alpha^2 = \beta^3 = 1, \quad \alpha\beta = \beta^2\alpha = \beta^{-1}\alpha.$$

The group has 6 elements and has presentation $\langle \alpha, \beta \mid \alpha^2 = \beta^3 = \alpha\beta\alpha^{-1}\beta = 1 \rangle$.

Chapter 9

9.1. Since $g^{-1}Ng = N$ for all g in G, it is certainly the case that $g^{-1}Ng = N$ for all g in H. So $N \lhd H$.

9.2. Let $a \in H \cap N_1$ and $b \in H \cap N_2$. Then $b^{-1}ab \in N_1$ since $N_1 \lhd N_2$, and $b^{-1}ab \in H$, since $a, b \in H$. Hence $b^{-1}ab \in H \cap N_1$, and so $H \cap N_1 \lhd H \cap N_2$.

9.3. The group encountered in Section 7.7 provides an example. In the notation of the example in that section, we have $H_5 \lhd H_2$, $H_2 \lhd G$, but H_5 is not normal in G.

9.4. One way round this is clear, since cyclic groups are certainly abelian. So suppose, in the usual notation, that G_{i+1}/G_i is abelian. Certainly G_{i+1}/G_i is soluble, by Corollary 9.7, and so there exist subgroups $H_0 = \{1\} \lhd H_1 \lhd \cdots \lhd H_k = G_{i+1}/G_i$ such that H_{j+1}/H_j $(j = 1, 2, \ldots, m-1)$ is cyclic. It follows from Exercise 1.31 that there exist subgroups $K_0 = G_i \lhd K_1 \lhd \cdots \lhd K_m = G_{i+1}$ such that $K_{j+1}/K_j \simeq H_{j+1}/H_j$ for all j.. If we do this for each G_{i+1}/G_i, we obtain an extended sequence of subgroups in which all the quotients are cyclic.

9.5. Write $a \sim b$ to mean "a is conjugate to b"; that is, if there exists x in G such that $x^{-1}ax = b$. Then $a \sim a$ for every a, since $e^{-1}ae = a$ ($\sim$ is *reflexive*). Next, if $a \sim b$, then it follows that $b \sim a$, since $(x^{-1})^{-1}bx^{-1} = a$ ($\sim$ is *symmetric*). Finally, if $a \sim b$ and $b \sim c$, so that $x^{-1}ax = b$ and $y^{-1}by = c$, then $(xy)^{-1}a(xy) = c$, and so $a \sim c$ ($\sim$ is *transitive*).

9.6. If $ga = ag$ and $ha = ah$, then $(gh)a = gah = a(gh)$, and so $gh \in Z(a)$. Also, from $ga = ag$ it follows that $ag^{-1} = g^{-1}(ga)g^{-1} = g^{-1}(ag)g^{-1} = g^{-1}a$, and so $g^{-1} \in Z(a)$.

9.7. (i) If $ax = xa$ and $bx = xb$ for all x in G, then $abx = axb = xab$, and so $ab \in Z$. Also $a^{-1}(ax)a^{-1} = a^{-1}(xa)a^{-1}$, and so $xa^{-1} = a^{-1}x$. Thus $a^{-1} \in Z$.

   (ii) Let $a \in H$ and $x \in G$. Then $x^{-1}ax = x^{-1}xa = a \in H$, since $H \subseteq Z$, and so H is normal.

   (iii) $a \in Z$ if and only if $x^{-1}ax = a$ for all x in G, that is, if and only if $C_a = \{a\}$.

Chapter 10

10.1. The polynomial $f = X^5 - 6X + 3$ is irreducible, by the Eisenstein criterion. From the table of values

X	-2	-1	0	1	2
f	-17	8	3	-2	11

we deduce that there are roots in the intervals $(-2, -1)$, $(0, 1)$ and $(1, 2)$. The zeros of the derivative f' are at $\pm \sqrt[4]{(6/5)}$, and $f'(X)$ is positive except between the zeros. Hence there are no other real roots and so, by Theorem 10.4, $f(X) = 0$ is not soluble by radicals.

10.2. The polynomial $f = X^5 - 4X + 2$ is irreducible by the Eisenstein criterion. From the table of values

X	-2	-1	0	1	2
f	-22	5	2	-1	26

we deduce that there are roots in the intervals $(-2, -1)$, $(0, 1)$ and $(1, 2)$. The zeros of the derivative f' are at $\pm \sqrt[4]{(4/5)}$, and $f'(X)$ is positive except between the zeros. Hence there are no other real roots and so, by Theorem 10.4, $f(X) = 0$ is not soluble by radicals.

10.3. (i) $\Rightarrow$ (ii). Suppose that $\{\alpha_1, \alpha_1, \ldots, \alpha_n\}$ is algebraically independent over K. If α_r were algebraic over $L_{r-1} = K(\alpha_1, \alpha_2, \ldots, \alpha_{r-1})$, it would have a minimum polynomial m in $L_{r-1}[X_r]$. If, for $i = 1, 2, \ldots, r-1$, we change each α_i in the coefficients of m to X_i, we obtain a non-zero polynomial $\overline{m}$ in $K[X_1, X_2, \ldots, X_r]$ such that $\overline{m}(\alpha_1, \alpha_2, \ldots, \alpha_r) = 0$. This is a contradiction.

(ii) $\Rightarrow$ (iii). Suppose inductively that

$$\sigma : K(X_1, X_2, \ldots, X_{r-1}) \to K(\alpha_1, \alpha_2, \ldots, \alpha_{r-1}),$$

given by $\sigma(f(X_1, X_2, \ldots, X_{r-1})) = f(\alpha_1, \alpha_2, \ldots, \alpha_{r-1})$, is an isomorphism. If α_r is transcendental over $K(\alpha_1, \alpha_2, \ldots, \alpha_{r-1}) = L_{r-1}$, then $L_{r-1}(\alpha) \simeq L_{r-1}(X_r) \simeq K(X_1, X_2, \ldots, X_{r-1})(X_r) = K(X_1, X_2, \ldots, X_r)$.

(iii) The equivalence of (i) and (iii) is essentially contained in the definitions.

10.4. $\mathbb{Q}(\alpha) \simeq \mathbb{Q}(X)$, and so by the argument of Theorem 3.16, is countable. The set of elements that are algebraic over $\mathbb{Q}(\alpha)$ is once again countable. Hence, since $\mathbb{R}$ is uncountable, there exists an element β in $\mathbb{R}$ that is transcendental over $\mathbb{Q}(\alpha)$.

10.5. After a bit of calculation,

$$t_1^3 + t_2^3 + t_3^3 = (t_1 + t_2 + t_3)^3 - 3(t_1 + t_2 + t_3)(t_1 t_2 + t_1 t_3 + t_2 t_3) + 3 t_1 t_2 t_3$$
$$= s_1^3 - 3 s_1 s_2 + 3 s_3.$$

Chapter 11

11.1. Π_n is constructible for $n \leq 100$ if and only if n is one of the numbers 3, 4, 5, 6, 8, 10, 12, 15, 16, 17, 20, 24, 30, 32, 34, 40, 48, 51, 60, 64, 68, 80, 85, 96.

Bibliography

[1] A. Baker, *Transcendental Number Theory*, Cambridge University Press, 1979.

[2] Carl B. Boyer, *A History of Mathematics*, Wiley, 1968.

[3] T. S. Blyth and E. F. Robertson, *Basic Linear Algebra*, 2nd Edition, Springer, 2002.

[4] E. T. Copson, *Functions of a Complex Variable*, Oxford University Press, 1935.

[5] Harold M. Edwards, *Galois Theory*, Springer, 1984.

[6] Paul R. Halmos, *Naive Set Theory*, Van Nostrand, 1960.

[7] John M. Howie, *Real Analysis*, Springer, 2001.

[8] John M. Howie, *Complex Analysis*, Springer, 2003.

[9] G. Karpilovsky, *Topics in Field Theory*, Elsevier, 1989.

[10] J. J. O'Connor and E. F. Robertson, *History of Mathematics Website, University of St Andrews*, (http://www-history.mcs.st-and.ac.uk/history/).

[11] H. Osada, The Galois group of the polynomials $X^n + aX^l + b$, *J. Number Theory* **25** (1987) 230–238.

[12] I. R. Shafarevich, Construction of fields of algebraic numbers with given solvable Galois group, *Trans. American Math. Soc.* (2), **4** (1956) 185–237.

[13] D. A. R. Wallace, *Groups, Rings and Fields*, Springer, 1998.

List of Symbols

Index